Arvind Sathi

Cognitive (Internet of) Things

Collaboration to Optimize Action

Arvind Sathi
IBM
Irvine, California, USA

ISBN 978-1-137-59465-5 ISBN 978-1-137-59466-2 (eBook)
DOI 10.1057/978-1-137-59466-2

Library of Congress Control Number: 2016953144

Cover illustration: Jacket image by sputnikos/iStock /Getty Images Plus

Printed on acid-free paper

This Palgrave Macmillan imprint is published by Springer Nature
The registered company is Nature America Inc. New York
The registered company address is: 1 New York Plaza, New York, NY 10004, U.S.A.

To my wife Neena, my chief historian

Acknowledgements

My first real introduction to artificial intelligence (AI) was in the Spring of 1983, when Mark Fox gave a talk to the graduate students at Carnegie Mellon University's (CMU) business school. Mark inspired me into imagining a future cognitive business organization and was well connected with business problems in manufacturing. I was searching for a Ph.D. research topic and eventually gathered the courage to walk up to his office at CMU's Robotics Institute and beg for a research position. He was at that time the director of Intelligent Systems Lab and was working on a number of research projects to explore a knowledge representation language named Schema Representation Language (SRL). The Robotics Institute and the Computer Science department provided me with a wonderful exposure to a grand vision of Cognitive Things—including vision, mobile robots, speech analysis, and factory automation. Over the next three years, I got a chance to meet and learn from some of CMU's masters in the field—Herb Simon, Raj Reddy, Mark Fox, Jaime Carbonell, John McDermott. Pat Langley, John Anderson, and many more. Mark also was the one to introduce me to Distributed Artificial Intelligence, which inspired the concept of Cognitive Things. At that time, I was dealing with Franz Lisp running on Unix machines with minimal interfaces to other agents. My discussions with Rick Rashid (who later founded Microsoft Research in 1991) and Alfred Specter (who built distributed engines for Transarc, and ended up working for IBM and Google) on distributed processing were invaluable in shaping my thoughts in this area. A number of workshops in Distributed AI provided me with key concepts represented in this book, and led me to valuable information exchanges with Victor Lesser, Reid Smith, and Tom Malone.

…ox asked me to teach a course on SRL at Digital Equipment in …d thus started my relationship with the Carnegie Group, an AI … formed with equity capital from Digital Equipment, GSI, Texas …uments, Ford Motor Company, Boeing, and U S WEST. I had the …portunity to work with the AI groups at each of these equity partners, and also to work with a number of AI scientists, who were brave enough to work on industrial applications—Gary Kahn, who worked on an early version of the Field Service advisor, Phil Hayes (now working at IBM Watson group), who worked on language translation, and Vijay Saraswat (now with IBM Research), who introduced me to Prolog. In 1987, *BusinessWeek* declared AI winter, but Carnegie Group kept building business applications of AI, and eventually went through an IPO in 1995. This was a valuable experience in learning how to commercialize AI, though it was ahead of its time.

Watching the buzz for Cognitive Computing today is like viewing the sleeping beauty (frozen by the curse of AI winter back in the 1980s) awakening. I am grateful to the IBM senior leadership for taking the bold move towards artificial intelligence, and facilitating the much-needed focus towards successful implementation. I would to thank Ravesh Lala for pulling me into his team and bringing my youth back. Swami Chandrasekaran and I have worked with each other since 2001, but never had so much fun with products and ideas. Many of the ideas represent our joint work and I would like to thank Ravesh and Swami for countless brainstorms. The Cognitive Value Assessment team at IBM brings the best of business and technology talent and has contributed to the use cases presented here. I would like to thank Grady Booch for his presentation, which helped me shape the definition. A number of my peers at IBM have helped me with this book. They include, among others, Aadhar Garg, Ayan Bhattacharya, Bart Emanuel, Bjorn Austraat, Brian Roumpz, Chris Loza, Christine Twiford, Dakshi Agrawal, Doug Barton, Eliana F Bennett, Eric Riddleberger, Janki Vora, Khalid Behairy, Manish Sampat, Manprit Singh, Mathews Thomas, Mikhail S Gorbatovskiy, Oliver Blodgett, Onye Ifemedebe, Owen Coelho, Rahul Kurane, Rajiv Phougat, Rich Lanahan, Sham Vaidya, Steve Trigg, Susan Visser, Tommy Eunice, and Shuvanker Ghosh. A number of IBM clients, educators, and business partners have helped me shape ideas, including Sofia Gomes, Anand Sekhar, Ash Kanagat, Dave Dunmire, Girish Varma, Jamie Williams, Joshua Koran, JP Benini, Manav Mishra, Maureen Little, Raj Singh, Sanjeev Dewan, Sumit Chowdhury, Vinodh Swaminathan, and Von McConnell. I would like to thank IBM senior leadership for all the publicly available media referred to in this book.

In my last two books, I sought help from third party artists to
cartoons using my ideas. My cartoons were not the best, though the
of using cartoons to convey some of the ideas was appealing. Fortu
came across Sunil Agarwal, a fellow IITian, who has been creating ca
for *Times of India*. I started subscribing to his email subscriptions, and fo
his cartoons were very much connected with technical topics. He has gra
ciously allowed me to share some of his cartoons through my work, and I have
included a number of his cartoons in this book.

I would like to thank Laurie Harting, Marcus Ballenger, and Steve Partridge for all their support in editorial review, and publication help, and to IBM Marketing—Will Reilly, Doug Barton, Susan Visser, and Gaurav Deshpand—for their help in accessing material for the book. I would like to acknowledge the timely and detailed reviews by Bjorn Austrat and Christine Twiford. My daughter Kinji Sathi is the most patient and detailed reviewer of my writing, and has provided valuable assistance in improving the writing quality. My son, Conal Sathi, provided a thorough technical review to keep me honest. I would like to thank my family, Oliver, Clark, Conal, Kevin, Kinji, Neena, and my dad for their support during the long months of book writing and for giving me valuable ideas for case studies.

Contents

1 **Introduction** 1
1.1 Background 1
1.2 What Are Cognitive Things and How Do They Function? 3
1.3 Book Outline 9
1.4 Target Audience 12
1.5 Summary and What's Next 12

2 **What Is a Cognitive Device?** 13
2.1 Background 13
2.2 Candidate Devices 14
2.3 Cognitive Device Operation 21
2.4 Cognitive Device Engineering 23
2.5 Blockchain for Transaction Management 25
2.6 Chapter Summary 26

3 **Cognitive Devices as Human Assistants** 29
3.1 Introduction 29
3.2 Shopping and Buying Assistant 31
3.3 Care Assistant 33
3.4 Travel and Entertainment Assistant 35
3.5 Administrative Assistant 37
3.6 Chapter Summary 39

…tive Things in an Organization 41

Introduction 41

Smarter Operation 44

…3 Smarter Engineering 49

4.4 Contextual Marketing 53

4.5 Proactive Customer Care 55

4.6 Counter Fraud Management 58

4.7 Chapter Summary 58

5 Reuse and Monetization 61

5.1 Introduction 61

5.2 Weather Services 62

5.3 Media Viewership Services 64

5.4 Location Services 66

5.5 Location Data and Contextual Marketing 68

5.6 The New Advertising Market Place 70

5.7 Monetization Candidates and Criteria 73

5.8 Cognitive Monetization 75

5.9 Chapter Summary 76

6 Intelligent Observations 79

6.1 Introduction 79

6.2 Sense 80

6.3 Observe 82

6.4 Listen 84

6.5 Crawl 86

6.6 Visual Recognition 87

6.7 Identity Resolution 88

6.8 Chapter Summary 90

7 Organization of Knowledge and Problem-Solving 93

7.1 Introduction 93

7.2 Organizing Solution Space 95

7.3 Text Analysis 99

7.4 Profile Enrichment 102

7.5 Automated Problem Solving 102

7.6 Adaptive Real-Time Decision-Making 105

7.7 Chapter Summary 108

8 Installation, Training, Maintenance, Security, and Infrastructure 111

8.1 Introduction 111
8.2 Installation and Maintenance 113
8.3 Training 114
8.4 Security and Privacy 116
8.5 Centralized or Distributed Architecture 119
8.6 Cloud or On-Premise Infrastructure 121
8.7 Chapter Summary 122

9 Machine-to-Machine Interfaces 125

9.1 Introduction 125
9.2 Communication Media 126
9.3 Identity Management 129
9.4 Information Governance 131
9.5 Negotiation 133
9.6 Chapter Summary 135

10 Man-to-Machine Interfaces 137

10.1 Introduction 137
10.2 Authentication 139
10.3 Cognitive Interaction 142
10.4 Emotions, Creativity, and Hidden Meanings 146
10.5 Negotiation/Disambiguation 148
10.6 Chapter Summary 149

11 Assisting in Human Communications 151

11.1 Introduction 151
11.2 Information Integration and Discovery 152
11.3 Alternative Generation and Prioritization 154
11.4 Conversation Assistance 156
11.5 Organization Communication 157
11.6 Chapter Summary 159

12 Balance of Power and Societal Impacts 161

12.1 Introduction 161
12.2 Displacement of Cognitive Jobs 163
12.3 Who Is the Winner? 166

12.4 Regulatory Versus Consumer Privacy 167
12.5 Changing Role of Machines and Humans in Families and Organizations 169
12.6 Organization Design, Policy Management, Change Management 171
12.7 New Skills and Shortage Areas 172
12.8 Chapter Summary 174

Index 177

List of Figures

Fig. 3.1	Ivy in action	37
Fig. 4.1	Ollie: The first driverless vehicle to integrate IBM Watson	52
Fig. 4.2	Etisalat location data monetization	54
Fig. 4.3	Proactive care scenario	56
Fig. 5.1	Advertising in the broadcasting era	71
Fig. 5.2	Today's advertising market place	71
Fig. 5.3	Advertising market participants	72
Fig. 7.1	Discover, detect, decide, and drive using advanced analytics platform	106
Cartoon 8.1	Dilbert's security (Scott Adams, "Dilbert", February 28, 2016, http:// dilbert.com/strip/2016-02-28, reprinted with permission)	117
Cartoon 8.2	Would you like to receive pop-up ads from your toaster?	118
Fig. 8.1	Collaboration across distributed sources	120
Fig. 10.1	CogniToys from Elemental Path	138
Cartoon 10.1	Remember the Password	139
Cartoon 10.2	Robot with consciousness (Scott Adams, "Dilbert", November 23, 2015, http://dilbert.com/strip/2015-11-23 reprinted with permission)	146
Fig. 10.2	Watson Personality Insights	148
Fig. 12.1	Cognitive employment trends	165
Cartoon 12.1	Role of parent vs. driverless car	170
Cartoon 12.2	Interaction skills and the next generation	174

List of Tables

Table 1.1	Cognitive operators	8
Table 4.1	Sample organizational use cases	43
Table 7.1	Knowledge Representation Layers	98
Table 7.2	Sample data dimensions and data set	103
Table 9.1	OSI layers	128
Table 10.1	Comparison of authentication technologies	142

1
Introduction

1.1 Background

I have always taken pride in an active life and a balanced diet. However, my belly finally reacted to the extended time in front of a computer screen and the nearby kitchen pantry stocked with delicious munchies. My daughter, an internal medicine doctor, was concerned. She asked me if I had a sizable decrease in exercising, and I gave my usual pitch about my regular exercise pattern. She was not satisfied with my response and gave me a FITBIT(R) as a Christmas gift. Within a week of my wearing the Fitbit, she was "taunting" me about my lack of exercise.

Exercising is a very personal activity, and yet it is highly influenced by our social network. People traditionally spend a lot of time making friends to exercise with, this network being typically limited to a geographical proximity. Fitness trackers have extended that concept to the wider social network, or to a worldwide level using Internet connectivity. Friends and family members can observer each other's exercise patterns and share comments with each other. Fitness trackers combine five capabilities to build a collaborative exercising behavior.

1. They use sensors to *collect* event data from me *automatically*. As long as I wear the Fitbit, it is collecting data. In doing so, it is reasonably accurate and very user friendly. Even my 93-year-old dad is able to use a Fitbit to collect his movement data.
2. The tracker itself is a small *low energy-consuming* device connected using a Bluetooth®. Once charged, it lasts for a long time and gives me a fair warning before running out of battery. It uses a gateway to poll the data into a central storage.

A. Sathi, *Cognitive (Internet of) Things*,
DOI 10.1057/978-1-137-59466-2_1

3. The *central storage collates data* about me and about my social network. It has the capability for storing raw data, aggregating it over time, comparing it to my social group and communicating the results to those in my network. Each of us may have different social networks. It provides me with information not only about people connected to me but also to their friends, thereby giving me an opportunity to make new exercise friends.
4. My *social network is able to collaborate* with me in my exercising by viewing shared exercise data and then in turn encouraging me through comments. Fitbit provides us with emoticons ("cheer" and "taunt") to share with each other. In the example above, my daughter was using a "taunt" that showed up as a SMS text on my cell phone.
5. I can utilize *professional consulting* offered by the software, which uses knowledge of medicine and fitness activities. Under Armour's Record™ app monitors my exercise, weight, calories burned, heart rate, and eating patterns and provides me with expert advice using IBM's Watson™.

Potentially, this information could be shared with my primary care physician who can use it for monitoring my health, my health insurance provider who can give me discounts for healthy living, and with sports marketers who can use it for targeting campaigns for sports shoes. Many employers are subsidizing the cost of acquiring a Fitbit, Apple watch or other monitoring devices, as they perceive the value in driving healthy living programs among their employees. Thus, fitness trackers are enabling both collaborative exercising and health monitoring. In doing so, these *Cognitive Things* are facilitating a number of information processing activities—sensing, data sharing, comparing, correlating, interpreting, advising, alerting—to enable and support the related human *collaboration to optimize action.*

The "Internet of Things" (IoT) represents a growing sophistication among devices. Examples of devices in the network of the IoTs include mobile handsets, refrigerators, cars, fitness trackers, watches, eBooks, vending machines, and parking meters, and the number of types of devices is likely to grow exponentially over the coming years. These devices are already gathering and communicating massive amounts of data about themselves, which is collated, curated, and harvested by a growing number of smart applications.

Just connecting a device to the Internet does not result in collaboration. The core theme of this book is the identification of cognitive behavior among IoTs. A network of Cognitive Things uses a new computing paradigm—namely cognitive computing—along with the power of the Internet and the data available from a collection of devices to forge new collaborations and create new applications never imagined before.

This chapter introduces the major propositions outlined in this book. It provides a definition of "Cognitive Things" and the scope of devices discussed in this book. It introduces the concept of cognitive computing and summarizes the chapters, different potential reader personas and the area of focus for each persona.

1.2 What Are Cognitive Things and How Do They Function?

We are on the threshold of a massive explosion of connected things. A McKinsey report projects the potential business impact as $4–11 trillion per year by 2025 using nine settings—factories (e.g., preventive maintenance), cities, human (e.g., improving wellness), retail, outside (e.g., self-driving vehicles), work sites, vehicles, homes, and offices.[1] There are many other projections each defining IoTs and projecting their impact in the trillions of dollars.

How do the IoTs derive such large business impacts? Let me illustrate an example to show how far reaching these IoTs will be in disrupting markets and businesses. *Business Week* has projected the availability of driverless cars to premium customers by 2025 and is also predicting driverless technology taking over taxi and ride-sharing fleets by 2030.[2] While the driverless car includes a large number of sensors to collect information about the car and the road, the autonomous vehicle is far more than a collection of sensors connected to the Internet. It is actually replacing the driver! Through automatic gear change, cruise control, automated lane detection, to automated parking, we have seen a number of ways in which vehicles are beginning to use this data to perform tasks originally performed by the drivers. We[3] will see a maturing technology capable of making a series of cognitive decisions to drive the car without requiring a driver. Instead of purchasing cars, potential riders may in the future

---

[1] James Manyika et al., "The Internet of Things: Mapping the Value Beyond the Hype", Executive Summary of the Report titled "Unlocking the Potential of the Internet of Things", McKinsey Global Institute, McKinesy & Company, June 2015, http://www.mckinsey.com/insights/business_technology/the_internet_of_things_the_value_of_digitizing_the_physical_world

[2] Keith Naughton, "Can Detroit Beat Google to the Self-Driving Car", http://www.bloomberg.com/features/2015-gm-super-cruise-driverless-car/

[3] I have used the term "we" to represent the humans, who are likely to be the consumers of the Cognitive Things.

use taxi services to move them from point A to point B, allowing for less public parking and potentially higher vehicular speed. These changes will have a profound impact on the auto insurance business, car dealers, taxi operations, rental cars, and public transportation. Moreover, they could easily translate to a sizable share of the $11.1 trillion business impact projected by the McKinsey study referenced above.

According to Dr. John E Kelly III, IBM's senior vice president, cognitive is the third era of computing. The first, tabulating, began in the late nineteenth century and enabled such advances as the ability to conduct a detailed national census and the United States' Social Security System. The next era, programmable computing, emerged in the 1940s and enabled everything from space exploration to the Internet. Cognitive systems are fundamentally different. Because they learn from their interactions with data and people, they continuously improve themselves. So cognitive systems never get outdated and only get smarter and more valuable with time. This is the most significant paradigm shift in the history of computing.[4]

Let me use an encounter to describe a Cognitive Thing. As I discuss the concepts underlying my book with a wide spectrum of people, I get interesting responses, ranging all the way from disbelief to personal encounters. The following story was recounted to me by Dan Abercrombie, CEO of Abletech. The story covers Dan's personal interaction with a robot developed by research work at the Osaka University.[5] While many of the features described here seem futuristic, it will likely be commonplace to find robots providing concierge services at hotels, conferences, and musical and sport events. These concierges will display a range of cognitive capabilities, including:

- natural language comprehension
- empathetic conversation
- facial recognition
- context retention and recall
- knowledge of organization, products, and people

I was walking the expo floor at Semicon, and saw the banner from the booth of the Fujikin Corporation. One of their executive staff, a Mr. Suzuki (fictitious

[4] John Kelley," Smart Machines: IBM's Watson and the Era of Cognitive Computing", Columbia Business School Publishing, September 2013, http://www.amazon.com/Smart-Machines-Cognitive-Computing-Publishing/dp/023116856X/

[5] ResOU, "Leonardo da Vinci comes back as an android", Research at Osaka University, August 21, 2015, http://resou.osaka-u.ac.jp/en/research/2015/20150821_1

name), is a long-term industry contact and friend. I had a few minutes and decided to try to say hello to Mr. Suzuki. I walked into the booth and asked one of the (human) booth staff (in Japanese, which I speak fluently) if Mr. Suzuki was there that day. To my surprise, a voice coming a couple of meters from my right said, "Suzuki-san desu ka? Saki made imashita yo." (Something like, "Oh, you would like to see Suzuki-san? He was here until a few minutes ago.") I turned around to see a distinguished gentlemen dressed as Leonardo Da Vinci sitting on a chair with a microphone. Upon somewhat closer inspection, it turned out that our Da Vinci was a chillingly lifelike robot.

Recovering from my initial surprise, I stepped in front of the robot and it greeted me cordially in English, and made comment about the booth being very busy and crowded, and that Mr. Suzuki would be back later. It was clear the robot had recognized my Caucasian appearance and decided that it should switch to English with me. Mildly miffed by Da Vinci's presumptuousness (how did it know I wasn't French?), I mischievously decided to switch to Japanese, saying "Sumimasen ga, eigo wo wasurete shimatta." ("Sorry, but I have forgotten how to speak English.") Da Vinci's rejoinder in a thick Osaka dialect came without hesitation, "Ah Ha Ha Ha Ha... Washi ha italia-go wasureta! O-taku nihonngo jouzu ya na"; something like "Ha, ha, ha...I have forgotten my Italian! Your Japanese is really great!"

By this point, I was totally off guard. In less than thirty seconds of interaction, the machine had overheard my initial inquiry, appropriately responded faster than any of the five or six humans in the vicinity, adjusted its choice of language based on my appearance and then switched languages again when requested, making a natural and spontaneous joke at the same time. I bantered with the robot for another couple minutes and then went on my way. About five hours later, I strolled by the Fujikin booth and Da Vinci once again, and this time Da Vinci called out to me as I walked in the aisle, "Ah, mata irasshita, Suzuki-san ha asoko ni imasu yo." (Oh you are back. Suzuki-san is over there.) Suzuki-san, hearing his name mentioned, turned around in surprise, and saw me. We had a short conversation and well-wishes, both of us mixing in plenty of nervous laughter as Da Vinci continued to interject comments into our conversation, and somehow both of us feeling "obligated" to be polite to Da Vinci and explain that we were long-term business acquaintances and all. Somewhat uncomfortable with Da Vinci participating, I made my greetings with Suzuki-san cordial, but brief, and went on my way again, acutely conscious of the reality that the need to deal with thinking machines will become routine in my lifetime.

Robots are beginning to perform the cognitive functions depicted in this encounter—recognize people, chit-chat, apply conversational humor, recall context in a later conversation, and be aware of the environment and the presence of others. In Chapter 10, I will be introducing some of these cognitive functions in detail, and how they are utilized in human conversation with a Cognitive Thing.

What is cognitive computing, and how does it relate to artificial intelligence and expert systems? A cognitive system learns at scale, reasons with purpose and interacts with humans naturally.[6] "Cognitive Computing" refers to automated agents that can learn complex tasks, interact with humans via natural interfaces and make autonomous decisions and actions working with individual and groups. It represents a new generation of computing systems enabling genuine human–machine collaboration where the system is able to understand high-level objectives specified by humans in a natural language, autonomously learn how to achieve the objectives from data in the domain, report results back to humans, and iterate the interactions via sequential dialog until the objectives are achieved. To enable a natural interaction between them, cognitive computing systems use image and speech recognition as their eyes and ears to understand the world and interact more seamlessly with humans. It provides a feedback loop for machines and humans to learn from and teach one another. By using visual analytics and data visualization techniques, cognitive computers can display data in a visually compelling way that enlightens humans and helps them make decisions based on data. Cognitive computing systems get better over time as they build knowledge and learn a domain—its language and terminology, its processes and its preferred methods of interacting. Unlike expert systems of the past, which required rules to be hard coded into a system by a human expert, cognitive computers can process natural language and unstructured data and learn by experience, much in the same way humans do. While they will have deep domain expertise, instead of replacing human experts, cognitive computers will act as a decision support system and help them make better decisions based on the best available data, whether in healthcare, finance or customer service.[7]

The starting point for IBM's cognitive journey was the Watson system with its first major accomplishment of defeating human *Jeopardy* experts using a computer program. While it was a great accomplishment, *Jeopardy* actually represented a relatively simple cognitive activity, as it is a single question represented in the form of an answer and requires search in unstructured content to find an answer and to pose it back as a question. Most cognitive tasks are far more complex, and in many cases it is extremely difficult, if not impos-

---

[6] John E Kelly, "Computing, cognition, and the future of knowing", IBM Research, October 2015, http://www.research.ibm.com/software/IBMResearch/multimedia/Computing_Cognition_WhitePaper.pdf

[7] Biplav Srivastava, Janusz Marecki, Gerald Tesauro, "2nd Workshop on Cognitive Computing and Applications for Augmented Human Intelligence, in conjunction with International Joint Conference on Artificial Intelligence (IJCAI)", Buenos Aires, Argentina, July 25-31, 2015, https://sites.google.com/site/cognitivecomputing2015/

sible, to replicate human function. Cognitive Things tend to embody selective cognitive functions to either augment or in some simple cases replace human activities. In a presentation to IBM's Academy of Technology, Grady Booch, well-known co-author of *The Unified Modeling Language User Guide* and now an IBM Fellow, introduced his definition of *embodied cognition*:

> Imagine unleashing Watson in the physical world. Give it eyes, ears, and touch, then let it act in that world with hands and feet and a face, not just as an action of force but also as an action of influence. This is embodied cognition: by placing the cognitive power of Watson in a robot, in an avatar, an object in your hand, or even in the walls of an operating room, conference room, or spacecraft, we take Watson's ability to understand and reason and draw it closer to the natural ways in which humans live and work. In so doing, we augment individual human senses and abilities, giving Watson the ability to see a patient's complete medical condition, feel the flow of a supply chain, or drive a factory like a maestro before an orchestra.[8]

So, how are automobiles getting transformed using cognitive computing? Driving is a cognitive activity. If the driver falls asleep or is impaired, the brain stops a large number of cognitive actions, each of which could be fatal to the car and the driver. A driver uses vision to recognize stationary obstacles, other moving objects, and the curvature of the road. The driver also has the ability to use a combination of steering, breaking and acceleration actions to control the movement of the automobile. The driver keeps track of fuel consumption and equipment failures (for example, a burst tire) and makes decisions. In a typical driverless car, the driver can go to sleep, get impaired, watch a movie or conduct a meeting, while the car provides all these cognitive functions. In addition, the data collected by the car can now be used for equipment failure prediction, improved engineering, or can be shared with city planners, who will use this data to optimize traffic flows. In such a case, a driverless car and its collection of sensors have the ability to collect the data automatically. Each sensor uses the car's energy to operate, and the data can be shared with car components. Also, an OBDII (On-Board Diagnostics) port provides external access to the data. With the help of an OBDII port and Bluetooth or Wi-Fi technology, this data can now be shared with a number of organizations, including the car manufacturer, parts supplier, auto service infrastructure, insurance companies, city planners, etc.

Table 1.1 compares cognitive operators to show how computers are evolving from the automation era to the cognitive era. They use evidence-based

[8] Grady Booch, presentation given to IBM's Academy of Technology, March 23, 2016.

Table 1.1 Cognitive operators

Function	Automation era	Cognitive era
Agent Interaction	Present facts, ask questions, retrieve responses, help, set default	Empathize, chit-chat, switch context, converse, negotiate, disambiguate, refine and change decisions, discuss and resolve problems
Solution Development	Compute, apply statistics, use decision tree, aggregate, search for key words, sort, select, delete, insert, join	Analyze (impact, sensitivity), reason, design a solution (synthesize, configure, extrapolate, discover), filter, focus, learn (inductive, deductive, deep, knowledge-based), context-sensitive search, decide, plan, schedule, diagnose, interpret, abstract
Event Ingestion	Extract, transform, load, stream	Sense, observe, listen, crawl, recognize, identify, reconfigure

reasoning. They seek to find new insights and connections from within vast amounts of information. They change the way humans and systems interact. Ingestion technologies in the automation era were geared towards large-scale data transfers focusing on extraction, transformation, and loading of the data to the target platform. In the Cognitive Era, the focus is on their ability to observe, recognize, and identify. In the case of the driverless car, the car must identify moving objects, especially those which are likely to collide with the car. The car must sense a malfunctioning car wheel or a rough terrain, and differentiate it from regular conditions. Information processing in the Cognitive Era moves computing technologies towards out-of-box approaches to problem-solving, such as synthesizing a solution, which is a much harder problem as compared to choosing an alternative based on static decision trees, an automation era computing capability. Some of the related planning activities are already available to drivers today—such as dynamically rerouting based on changing conditions. Last but not least, the agent interactions are far more human-like with the ability to empathize and negotiate. Will a driverless car yield to another car while changing lanes or decide how to reduce the amount of damage in an unavoidable accident, choosing saving lives over wrecking another driverless car without passengers?

Table 1.1 represents many cognitive operators. As in the case of humans, the cognitive thing does not need to exhibit all of these operators to pass the "cognitive" test. Is there a cognitive IQ for the machine? This is a cognitive journey and it will take many decades to mature to the level where Cognitive Things will exhibit a majority of these operators in production systems. The race today is in finding a couple of nuggets that provide the biggest business value and use them to initiate the journey. As will be shown through the examples in this book, the key to success is in striking the right balance between man and machine to achieve optimal results in the short term, and adding additional capabilities in later phases, as dictated by their business value and appropriateness.

1.3 Book Outline

In literature and the media, we have seen many futuristic examples of smart refrigerators and coffee machines and how they will simplify our daily life. What is our expectation of the Cognitive Things and how pervasive is it likely to get? Certainly, we are used to a large number of not-so-smart machines capable of supporting our day-to-day activities with an increasing level of automation. In the Cognitive Era, an intelligent home could be equipped with smart televisions, smart refrigerators, smart dish washers, smartphones, smart cars, and so on. In the automation era, these devices were primarily designed for automating the tasks in mechanical ways. As these machines become intelligent, they must learn how to listen, reason, perceive and change with our wishes. Each of these devices must be configured and trained to deal with their human counterparts. Do I like my coffee in the early morning or mid-morning? Do I take it with cream and sugar? If I have to engage in training my coffee machine each time a guest arrives and alters its daily routine, and retrain when the guest departs, where is the value to me as a consumer? Why should I use this "smart" feature and pay additional money? Is there a business use case for a cognitive coffee machine, which can sense my preferences and auto-configure itself, and is that really feasible in the near future?

This book examines Cognitive Things, covering a number of important questions:

- What are Cognitive Things?
- How do they interact with other machines and with us?
- Can machines support human communication as intelligent listeners or aids to procure supporting facts, or evaluate alternatives?
- Which technical components make up cognitive behavior?
- What applications can be driven from Cognitive Things—today and tomorrow?
- How does it redistribute the workload between humans and machines?
- What types of data can be collected from them and shared with external organizations?

- How do they recognize and authenticate authorized users? How is the data safeguarded from potential theft? Who owns the data and how are data ownership rights enforced?
- How do these devices collect and share data about themselves, how is this data collated, secured, obfuscated, and how can this data be harvested by smarter applications to provide new innovative capabilities for a variety of users?
- What insights can be generated not only about Cognitive Things but also about the people who interact with them?
- Does it change current corporate infrastructures for organizations?

This book covers three related directions associated with Cognitive Things: business use cases, technical capabilities, and impact on consumers and corporations. The business use cases describe how the Cognitive Things will be used and justified in their value propositions. The technical capabilities discussion explores the technical feasibility of engineering these solutions. The impact on humans and organizations explores the integration aspects in today's world and how individuals and organizations are likely to adopt Cognitive Things.

The first part of the book explores compelling business use cases to drive Cognitive Things. Do these use cases relate to objects or humans, individuals or organizations, inside an organization or in a marketplace? Chapter 2 identifies how Cognitive Things take care of themselves, in traditional maintenance, operations, engineering, and reconfiguration aspects. A series of common examples are used to describe the business value from a cognitive thing and the business capabilities covered. Chapter 3 identifies how Cognitive Things improve support for the individual customer. It examines new roles and responsibilities for the Cognitive Things and the value provided to the individual. Chapter 4 identifies how Cognitive Things support an organization. Chapter 5 covers how Cognitive Things are adding new data sources to the services marketplace and how organizations are using new business models to serve their business customers using monetization of data from Cognitive Things.

The second part explores the technical dimension identifying the key technical capabilities, which are required to fulfill these use cases. For each technical capability, the book outlines the level of sophistication required and available as of today. The data represents extreme velocity, volume, and variety, and most of this data is unstructured. It is impossible for any computing environment to collect, correlate, analyze all the data being generated, and act upon the results in near real-time. Additionally, appropriate tools and techniques can isolate micro-segments and learn finer details relating to these micro-segments. Advanced analytics techniques can help identify trends, focus on a tiny fraction of the data, enquire and collect additional data wherever needed, and decide how to act upon a situation. Cognitive Things will also speed decision-making through automated data collection and decision-making. Early detection and correction can result in substantial benefits to the individuals, organizations, and society.

This section is divided into four chapters. Chapter 6 focuses on data acquisition aspects and how Cognitive Things are able to acquire data. It examines the differences provided by the cognitive function and the capabilities required to realize it. Chapter 7 focuses on information processes and examines how new cognitive information processing functions use learning to provide better insight and more focused action. Chapter 8 covers maintenance and support aspects and how Cognitive Things can be more easily installed, configured, monitored, trained, and secured without requiring an army of maintenance professionals. It covers thorny issues around data protection, and rights management. While the book touches upon technical topics, it does not provide a detailed explanation of the underlying technologies. A number of books have been published on the technical aspects, and the respective chapters provide some useful references to the excellent material available elsewhere.

The third part of the book explores human and machine interfaces. It looks at how these changes will disrupt and change our current organizations and how can we prepare ourselves. It addresses some of the pitfalls that must be avoided. Chapter 9 covers machine-to-machine interfaces and how Cognitive Things can relate to other intelligent agents in conducting their task, supporting their customers and negotiating a better result for their owners. Chapter 10 covers human-to-machine interfaces and how cognitive devices can act as assistants for humans. Chapter 11 covers human-to-human communication with the use of cognitive devices as observers and advisers. Chapter 12 covers impact on organizations and society. It explores the opportunities and what must be done to integrate them with today's organizations. It also considers the nature of the man-machine relationship and how it changes with the Cognitive Things. It projects the changing organization structures and how the twenty-five billion Cognitive Things will change the way eight billion humans organize and run their daily lives, organizations and societies.

The book applies a series of public case studies to illustrate its assertions. Innovators and early adapters are already benefiting from the initial explosion of devices and capabilities available today. However, what you see today is only the tip of the iceberg. Using a series of futuristic visions collated throughout my research, the book additionally explores what is the new act of impossible and where we are headed.

1.4 Target Audience

The book is primarily targeted towards organizations marketing, manufacturing and operating a network of Internet of Things today or planning to do so in the future. Most of these organizations are engaged in activities to apply cognitive computing and make their applications smarter. The book will provide much needed information about potential use cases, business and IT capabilities, and ways of mitigating failure. The book is written with the business user in mind. It will cover a number of technical topics; however, it focuses on those topics from a business perspective, describing how these technical topics impact business issues, rather than diving into details of the technologies. Universities can use this book as reading material for their Information System courses. I will be creating additional material for educators interested in teaching cognitive computing courses.

1.5 Summary and What's Next

In this chapter, I have laid out the purpose and abstract of this book. Cognitive Things are capable of providing a number of key capabilities needed for collaboration including their ability to collect data automatically, data sharing across a network, and information processing for decision optimization.

The information processing is done cognitively and reflected in a new set of human-like capabilities in event ingestion, solution development, and agent-interface. In a typical domain such as a driverless car, these are necessary capabilities for a machine to think and act in a human-like way.

The key propositions included in the book are:

1. Business use cases are driving Cognitive Things and how they support the device, the individual, the organization and the external marketplace.
2. Technical capabilities have emerged to enable cognitive capabilities to be added to IoTs providing the Cognitive Things with the power to process information like humans.
3. The Cognitive Things will have enormous impact on consumers, and organizations, creating new ways to deal with collaborative activities and designing new policies.

In the next chapter, I will discuss how Cognitive Things function and how they support maintenance, operations, design, and engineering tasks.

2

What Is a Cognitive Device?

2.1 Background

When my wife and I relocated from Denver to Southern California, we decided to trash our 20-year-old bulky television and shop around for one of the latest models. Neither of us was in the habit of sitting and watching network television—we were far more interested in watching shows suited to our interest and convenience. We were also interested in purchasing cognitive appliances for our house and mistakenly thought a "smart tv" was likely to be a cognitive device. A number of television manufacturers were offering what they proclaim to be smart TVs. As we dived into the specifications of smart TV, we found to our astonishment, that so-called smart TVs were not so smart. They had no idea who was watching the television. They could not remember what I watched last. Worst of all, when the smart TV was not working, it could not self-troubleshoot among the device, Internet connectivity, and the content provider. Instead of taking care of its content delivery problem, it forced me to repair the television, integrate knowledge from many diverse areas, and converse with multiple technicians. A set of customer service professionals guided me to perform random acts of hot and cold reboots, hoping that those reboots would remove the underlying problem. The experts agree that televisions have a long way to go before they become cognitively smart.[1] While there are many ways a television today is far more sophisticated than

[1] Michael Miller, "Smart TVs: Viewing in a Connected World", Que, April 2, 2015. http://www.quepublishing.com/articles/article.aspx?p=2320941&seqNum=7

A. Sathi, *Cognitive (Internet of) Things*,
DOI 10.1057/978-1-137-59466-2_2

our clunky 1995 television and offers a variety of new programming using its Internet connectivity, the advances have very limited cognitive characteristics.

At the same time, I found a number of audience services, such as Netflix, which offered many cognitive characteristics. They reorganized the menus based on my viewing patterns, kept track of my wish list, offered recommendations, and differentiated among family members in the same viewing plan. Imagine a cognitive television, one that would be more like a server in a restaurant, that could relate to the person watching the television, serve based on content preferences, and take care of network issues instead of leaving all the troubleshooting to the owner. Like a good restaurant, the television would re-engineer itself around the most popular functions, getting rid of scarcely used menu items, simplifying the interface, making it easier to install, operate, and troubleshoot. Last but not least, the cognitive television must be secured from external hacks. Today's smart TV with a webcam is a handy device for a thief to check into a house before burglary. In a presentation to IBM, Collin Dunphy and Evan Spisak, aged 12 and 10 respectively, demonstrate hacks to take over household webcams for unintended access.[2]

The following sections identify the characteristics of a cognitive device and its functions, as compared to a non-cognitive device. While many of the features defined here are in production use somewhere on some device, the widespread application of this use case is still a (cognitive) thing of the future. However, it offers tremendous opportunity for the manufacturers to grab a market by offering the first truly cognitive device.

2.2 Candidate Devices

Let us look at a few examples to illustrate the behavior of a cognitive device. Many home appliances, such as washing machines, dishwashers, and security systems, are becoming cognitive. My daughter and son-in-law have set up their home to be a cognitive house connecting lamps and video cameras to the Internet. The video cameras provide them with a useful baby monitor, giving them full access to the movement of their twin infants conveniently from an app on their cell phones. The cameras only record the video when there is motion or sound, thereby filtering out all the less important static view of sleeping babies in their cribs. In this way, the camera is acting like the night watchman who keeps track of all activities, but only needs to report events that cross certain thresholds. By time-stamping the videos, it provides them with an accurate account of the children's sleeping behavior, playtime and feeding patterns.

[2] IBM Academy of Technology Channel, "Hack: Homes are compromisable by kids", https://youtu.be/3Q6rLQxgeLQ

With newborn twins, it is important to be able to track the babies' behaviors. My daughter and son-in-law configured the app Baby Connect for all of their caregivers, to record the activities for each baby, including feeds, sleep and bowel moments. Baby Connect appeals to parents who have enough time and energy to record the length of their child's naps or number of diaper changes.[3] Baby Connect replaces a caregiver's handwritten notes with a collaboration engine that allows parents and child care centers to communicate easily. As soon as an event is saved, it is immediately synchronized on each parent and caregiver account. Everybody has access to the information in real time. You can not only record feedings, nursing, naps, diapers, milestones and pumping, but also the baby's mood, temperature, what kind of game he's playing and his GPS location. You can also attach pictures.[4]

Most new household appliances come with Internet connectivity and can detect usage and troubleshooting patterns for preventive maintenance. Whirlpool is now manufacturing washing machines that incorporate their innovative CareSync™ System.[5] Your washer and dryer can easily connect to any mobile device, putting control and exceptional fabric care right at your fingertips.[6] These machines can track detergent use, predict defects and schedule repair. The valuable usage and performance data can now be fed to the engineering systems for improvements in product engineering. In the future, washing machines will become more cognitive, scheduling their workload at a time when other household water usage is minimal, self-diagnosing and scheduling preventive maintenance thereby avoiding major failures and reorganizing their menu options based on how they are being used.

How smart are the appliances and security systems in my house, and are they cognitive? Dogs have provided basic cognitive functions in household security for centuries. A well-trained dog knows how to keep an intruder away with a fierce bark but equally is welcoming to members of the household and frequent guests. Service dogs go a step further in providing valuable company,

---

[3] Bob Tedeschi, "Devoting Attention to a Child and a Phone, All at Once", New York Times, April 27, 2011, http://www.nytimes.com/2011/04/28/technology/personaltech/28smart.html

[4] BabyConnect, "Collaborative baby care using baby connect", YouTube, May 7, 2012, https://youtu.be/AYLApsUDMM0

[5] Whirlpool® USA, "Whirlpool Smart Top Load Connected Laundry Pair, October 23, 2015, https://youtu.be/2BGcGdNcSOk

[6] From Whirlpool's corporate website, http://www.whirlpool.com/smart-appliances/smart-top-load-washer-dryer/

guiding their masters and alerting others in case of medical attention.[7] My 93-year-old father lives with us and we were concerned about his health and how to keep watch over him as he was showing signs of health issues that may require medical attention. Since my wife is afraid of dogs, getting a service dog was not an option, and we needed an option that could alert the 911 services, if needed. I called my home security company and a call center sales person promptly offered a device my father could wear all the time. The device would alert emergency services when he presses a button or falls down. As I marveled through the sunny day scenario (the typical sales use case presented by the salesperson), my engineering mind was searching for rainy day scenarios (how the system could malfunction). I asked how the emergency services would enter the house if my father were unable to respond to the doorbell. The salesperson told me they would do whatever they had to, even if required breaking down the front door to get into the house. She also said, "If the house is wired with an electronic lock and the emergency services has the combination code, they will use it, but it is equally probable that they may end up breaking down the front door." With some rigorous training, I could train my service dog to recognize the difference between a normal and emergency situation and differentiate between an unwanted intruder and an emergency care person. A cognitive elderly care and security system should be able to do all of this without breaking down my front door every time my dad accidently drops the dongle on the floor. The interesting integration point that she ignored was that the security system has full control of the house and can welcome an emergency care professional while keeping intruders away. It seems like security systems are headed in the right direction, but for now we need both service dogs as well as monitoring services to get the desired features.

In Chapter 1, I introduced cars as cognitive devices. Cadillac is introducing its new car with Super Cruise, the company's most ambitious technology foray since automatic transmission. Pairing adaptive cruise control with lane-centering technology, it will allow drivers (or whatever they are called in the future) to let the car take over only while on the highway. Google's latest prototypes are already driving around Silicon Valley, where they are known as Koala cars because of their bulbous shape.[8] The cognitive components have added a number of new sensors to the car's growing list of data sources, including front, side and back mirrors. They are also equipped with plenty of pro-

---

[7] Morieka Johnson, "5 things you don't know about service dogs", MNN.com, http://www.cnn.com/2012/06/15/living/service-dogs-mnn/

[8] Keith Naughton, "Can Detroit Beat Google to the Self-Driving Car? Inside GM's fight to get to the future first", Oct 29, 2015, Bloomberg BusinessWeek, http://www.bloomberg.com/features/2015-gm-super-cruise-driverless-car/

cessors resembling a large network of computers. The cars have been offering an OBDII port for more than a decade, which provides a gateway to all the sensor data from the various car components. This port can be connected via Wi-Fi or Bluetooth to an Internet connection point. Its data sharing enhances the cognitive performance of the car. All the relevant data is transmitted to a central computer and can be utilized in a variety of ways, including preventive maintenance, engineering, and operations. Cars can already follow the car ahead of itself in a lane. Drivers use a variety of primitive techniques to communicate with other drivers, including the car horn, which, if used extensively, may create enormous noise pollution as can be observed in the busy roads of Delhi. Cognitive cars should be able to electronically communicate with other cars without the use of light or sound and negotiate with each other for the right of way. What happens to all the rash drivers who are rushing to their appointments and are making up for the delay by driving faster than the speed limit? Imagine the roads of tomorrow where a series of driverless cars will be driving at the speed limit, in an orderly fashion with no need to watch for tailgaters or reckless drivers. As cars become more sophisticated, we as consumers have less and less ability to diagnose their problems. However, cognitive cars should be able to self-diagnose and schedule their own maintenance, obviously keeping in mind their owners' convenience and preferences. In order to perform these functions, cars are starting to have the ability to organize their sensors, and to use the sensors to predict maintenance. I will discuss preventive maintenance capability in more detail later in this chapter.

Smartphones and tablets were the first natural Cognitive Things. They have plenty of capabilities to locally process data, Internet connectivity to connect globally for voice, data, and video communication, and an ability to collaborate with other devices and central servers. As a cognitive integration point for other IoTs, they are increasingly performing three sets of tasks. First, they are emerging as the universal user interface for other IoTs. As we described in the Whirlpool example above, the controls for smart devices can be managed via smartphones and tablets. Second, they are the natural connection hub for connecting other IoTs to the Internet. Using Bluetooth or other technologies, they can collate data from other devices, buffer as needed, and upload to central servers. Third, they can be used as a payment hub by providing the mobile wallet capability. For performing these functions, they must remain ON all the time, and should not malfunction. The preventive maintenance is driven by their ability to collect data about their performance, including temperature, battery consumption, data upload/download, and many other parameters. What is their ability to perform preventive maintenance? One of my smartphones was defective. It would randomly heat up, and its battery would discharge suddenly in 10 minutes or less. I found it extremely inconvenient, especially on the day I was running late to the airport with an electronic boarding pass on my cell phone. When I tried to get it fixed, the statistics collected were not sufficiently detailed enough to provide the root cause. The repair shop decided to play it safe and gave me a replacement phone. However, it is not clear what the problem was and whether a factory reset would have removed the problem. In many cases the problem may not be with the smartphones but the apps running on the phone or the network connecting those phones. Despite replacing the phone, the consumer may still observe the same issues. Telecom service providers often refer to this problem as "No Fault Found" referring to the pile of perfectly good smartphones returned by disgruntled customers. The returns are both expensive for the service provider, as well as the leading cause of Net Promoter Score reduction among customers. A cognitive smartphone should be able to alert its service provider as well as its owner about potential problems and find ways to self-repair. A collaborating collection of Cognitive Things representing the smartphones and the network elements—wireless antennas, network routers, switches, and so on—can observe faults as they build up and help troubleshoot problems before the consumer notices the problem.

"Wearables" are the next set of Cognitive Things. Often, as in the case of smart watches, they provide the mechanism for collecting physical measurements from their owners. Wearables often have smaller energy consumption requirements and use Bluetooth and gateways to connect to the Internet. For example, the Fitbit described in Chapter 1 collects movement information and uses Bluetooth connectivity to transmit movement information to a central server, where a social network can observe and collaborate on exercising activities. The HealthBox™ from Under Armour® is a kit that includes a chest strap for heart rate monitoring, a wristband, and a circular scale. The fitness wristband has a unique design that makes it comfortable to wear all day, even while sleeping. You have the option to synchronize the scale with the corresponding apps so that it remembers who you are each time you step onto it. The scale can also measure body fat percentage and keep track of weight loss goals. All of the items in the kit will communicate with the Record app. Under Armour has also released its first smart shoe. It collects movement data so wearers will not need another device to track workouts – the shoe does it itself.[9]

Clothing with sensor technology is still nascent, but major apparel brands like Under Armour, Ralph Lauren® and Levi's® have been working to develop their offerings. "The structure of textiles is the same as the structure of touchscreens

[9] Ahiza Garcia and Hope King, "Under Armour reveals 'HealthBox' 24/7 activity monitor", CNN Money, Jan 5, 2016, http://money.cnn.com/2016/01/05/news/companies/under-armour-healthbox-ces-2016/

which we're using in everyday mobile devices and tablets," said Ivan Poupyrev, founder of Google's Project Jacquard (which Levi's is a part of). A new video shows a biker wearing a jacket and communicating with it.[10] "This means that if you just replace some of the threads in the textiles with conductive threads, you should be able to weave a textile that can recognize a variety of simple touch gestures, like any touch panel you have on a mobile phone."[11]

Let us move on to a set of Cognitive Things applied "in the air". Amazon CEO Jeff Bezos had a big surprise for CBS correspondent Charlie Rose in late 2013. After their *60 Minutes* interview, Bezos walked Rose into a mystery room at the Amazon offices and revealed a secret R&D project: "Octocopter" drones that will fly packages directly to your doorstep within 30 minutes. It's an audacious plan that Bezos says requires more safety testing and FAA approvals, but he estimated that delivery-by-drone, called Amazon "Prime Air," will be available to customers in as soon as four to five years. Two years later, Amazon has released a version 2 and is now testing its safety with FAA with support from NASA.[12,13] As much as driverless cars respect the laws and protocols of a busy highway, drones must learn to interact with human environment around them. Like the UPS delivery person, they must navigate to an address, avoid getting chased by the dogs, and be gentle to curious kids. In the video released by Amazon, the customer will receive a notification from Amazon stating the package arrival time and they will be placing an Amazon marker to pin point where the drone will land. Drones will also be a new data source for valuable environmental information. Their bigger counterparts, commercial aircraft, are already collecting a vast amount of data about themselves. They are providing that data to the airlines, aircraft manufacturers, and the weather company.

Internet enablement has seen a lot more maturity in telecommunications networks where customers are demanding premium network and service quality, and there is fierce competition among service providers to shine in front of their customers. After all, customers gravitate towards communications service providers who can provide them with five bars of connectivity at

---

[10] Levi's, "Levi's® Commuter x Jacquard™ by Google Truck Jacket", YouTube, May 20 2016, https://youtu.be/yJ-lcdMfziwer

[11] Kristina Monllos, "5 Ways Marketers Are Already Putting Sensors to Work Next-gen ideas are sure to be the talk of CES", Adweek, Jan 2, 2015, http://www.adweek.com/news/technology/5-ways-marketers-are-already-putting-sensors-work-168777

[12] CBS News, "Amazon unveils futuristic plan: Delivery by drone", Dec 1, 2013, http://www.cbsnews.com/news/amazon-unveils-futuristic-plan-delivery-by-drone/

[13] Trever, Mogg, "Tonight: Jeremy Clarkson presents the Amazon Prime Air delivery drone", Nov 29, 2015, http://www.digitaltrends.com/cool-tech/amazon-prime-air-delivery-drone/

all their major hangouts. Service providers achieve that objective by monitoring their network equipment and using this information to improve service quality. As communications service providers migrate to smartphones, third party apps, and data services, increasingly the service quality is dependent on a number of factors outside of their control. Telecommunication networks are rapidly gaining cognitive skills in reporting their performance, spotting performance issues, reconfiguring to meet changing demand patterns, and offering insights to service and equipment providers for better design. As they have progressed in their cognitive journey, they have helped generate many important technical innovations, which can now be adopted by other industries. The telecommunication networks contain many intelligent devices, which are all connected and can provide a lot of data about their performance. A simple access of a Facebook page on a smartphone may generate hundreds of network events. These events must be collected, correlated, aggregated, and scored close to data sources to provide meaningful insight to downstream monitoring systems.

There are many other networks of intelligent devices in the business world. This includes smart equipment on the factory floor, corporate fleets of trucks for a distribution company, or smart grids in a utilities company. As the cost of Internet connectivity goes down and analytics options become technically and financially viable, these networks are beginning to mature in their cognitive journey. For example, utility companies have invested in creating smart meters for their consumers' electricity consumption. In the past, meters were read once a month by utility company employees, who walked all over neighborhoods each day physically reading these meters. With automated metering systems, the utilities companies can read meters much more frequently. This gave rise to a number of applications, which exploit the demand distribution to optimize energy consumption and distribution.

Smarter cities are increasingly making their resources Internet friendly. For example, many cities now use Internet enabled parking meters where money can be deposited using credit cards or phone apps. Once the parking meters are connected, the parking meters can inform their availability to intelligent cars driving around on the streets, thereby reducing the time and frustration associated with finding parking. Smarter cities can also use the utilization data to reorganize city parking spaces. The traffic data collected from the cars can be utilized for a variety of planning and operational activities, including traffic signal coordination, rerouting during disasters, street repairs, and so forth.

2.3 Cognitive Device Operation

How do cognitive devices operate and how do they differ from their automation era counterparts? In this section, I use a set of cognitive functions to compare and contrast the cognitive journey. Let us start by looking at the fitness tracker example. Over five hundred years ago, Leonardo Da Vinci came up with the concept of a pedometer. Sketches from the famous artist showed a gear-driven device with a pendulum arm that moved in time with a person's steps. Fast-forward a few hundred years and pedometers now come in all shapes and sizes—mechanical, digital, or simply a smartphone app.[14] A Fitbit can track physical activity using a pedometer. While pedometers use steps as a measure to quantify physical activity, there are many other options, including heart rate monitors that measure changes in heart rate, and GPS-enabled devices that track distance and terrain. Each of these devices provides a measure of physical activity which works in most situations but fails to work universally. For example, my Fitbit was fairly accurate in measuring my steps on a regular walk or jog, but was inconsistent in measuring physical activity while smooth ballroom dancing, which requires gigantic steps (as in the Waltz), versus rhythm dancing (as in Cha Cha) where the steps are tiny. In the world of Fitbit, a step is a step, and it cannot differentiate between my gigantic side step in the Waltz, which consumes a lot more energy, and the short step in Cha Cha. GPS is very good in measuring outside activities that traverse distances, but fails to properly understand a rigorous round of badminton in an indoor court, for example. Heart rate monitors are probably far better than the other two, but fail in situations where heart rates increase without actual physical activity, say while watching a horror movie. In each case, these devices use algorithms to estimate physical activity from their raw measurements and may in some cases adjust for age and weight differences. Despite their shortcomings, these devices are able to observe physical activity with relative ease and share insights from these observations by converting raw measurements to physical activity and then communicating those converted measurements to information stores where they can be aggregated, analyzed, and shared. Comparatively, the classic non-cognitive treadmills provided a measure using the speed of movement and displayed the result to the person using the machine. If I got bored jogging on the treadmill and decided to terminate the jog, there were no taunts from my social group. Today, the Endomondo app on my iPhone tracks my location and performance during

[14]Technology, "How Pedometers work", Runtastics, March 24, 2015, https://www.runtastic.com/blog/en/technology/how-pedometers-work/

jogs and posts it automatically to my Facebook page. It compels me to meet or exceed my targets and share a good jog with my friends. By using the terrain information, it also projects the steepness of the jogging trail and computes calories burned while exercising.

How would a cognitive weighing scale work differently from the classic scales we have used for decades? I decided to test the Under Armour Scale which comes equipped with the Record app that I installed on my iPhone. It showed up in a nice packaging and instructed me to first install the app. As I powered on the scale, it used a Bluetooth connection with my iPhone to gather facts about me and greeted me by my first name. My wife saw this shiny scale in the bathroom and decided to use it to get her weight. The scale was perplexed, as the biometrics did not match "Arvind", and asked her, if, as a new user, she could introduce herself by installing the app first. While this is a simple recognition step, the fact that the scale uses biometrics—current weight and body fat percentage—to differentiate between us, my wife and I do not need to manually change the settings to tell the scale which of us is using it. As and when I stand to measure my weight, it correctly addresses me, monitors and sends my biometric information to my app—using its recognition to update the correct app connected with the person whose weight is being measured.

Recognition and user differentiation is the first step in learning. Cognitive Things learn from their usage, and continually change their interactions with their users. In Chapter 9, I will be discussing new toys—aptly named "Cognitoys™"—which are intended to be cognitive friends for children. Cognitoys learn from interactions and keep morphing based on the interests and cognitive development of the child. Unlike many other classic toys, which offer buttons for interaction, cognitive toys do something different each time they are used, thereby retaining their novelty and keeping the child interested in continuing to play with the toy. This learning can be a useful feature for any household equipment—the television, washing machine, washer, dryer, coffee machine, or security system. In each case, we have patterns of use, and these patterns can be learned. Cognitive Things must recognize different users through a variety of ways: by face, by touch, by voice, by fingerprints, and by usage, and apply past learning to customize their interactions.

When many relatives or friends show up and they travel together in more than one car, they use a variety of means to communicate with us, and they are always concerned about who will drive with whom, because the interactions are limited to only the people in a single car. Will cognitive cars change how we form a fleet of cars to travel together?[15] Maybe the cars can recognize each other and depending on the need, they can open up a session to share conversations or music across many cars. In doing so, they will need to communicate with each other. If they are sharing a communication channel, they can also identify each other's location and make it easier for their respective drivers to follow each other. Today, some of these functions can be performed by various apps on smartphones belonging to the car drivers or passengers. For example, using AmpMe, a fleet of cars can start a party and share music.[16] Intelligent machine-to-machine communication is a difficult topic and it requires a significant amount of cognitive skills to share goals, constraints, activities, and responses to questions. Today's cognitive machines are beginning to learn how to interact with each other in simple ways. I anticipate a fair amount of negotiation skills to be developed for driverless cars, as these cars will need to differentiate between friends, strangers, and enemies.

As robots start to interact with humans, they need to learn how to anticipate and respond to motion, such as handshake or a hug. Cognitive assistants must differentiate emotions and change their communication style based on emotional changes. Sensing and responding to emotion is an important cognitive skill. Classic automation was emotionless and programmatic. A cognitive assistant exhibits a smiley face or a cheerful voice, but also is able to change the conversation based on perceived tone, facial expression, or words used in the question. As a cognitive assistant senses anger, frustration, or sarcasm, it must adjust to the conversation.

2.4 Cognitive Device Engineering

The usage of a device can be collected and analyzed for valuable feedback to the engineering team. There may be components that fail more often and if properly designed, could significantly reduce the maintenance costs. There may be features that are used once in a blue moon and yet clutter the user interface making it complex for day-to-day operation. A device may be hard

[15] Rajit Johri, Jayanthi Rao, Hai Yu, and Hongwei Zhang, "Spatiotemporal Perspectives of Connected and Automated Vehicles: Applications in Wireless Networking", IEEE Intelligent Transportation Systems Magazine, Summer 2016, http://ieeexplore.ieee.org/document/7457387/

[16] Ellie Zolfagharifard, "The ultimate party phone: Free AmpMe app lets users link handsets to play music together as one giant speaker", Daily Mail, September 24, 2015, http://www.dailymail.co.uk/sciencetech/article-3248361/The-ultimate-party-phone-Free-AmpMe-app-lets-users-link-handsets-play-music-giant-speaker.html

to install and an installation wizard may significantly improve the installation headache. How often can the engineering organization collect and analyze this usage feedback to improve the design.

Quality Function Deployment (QFD) is a method to transform qualitative user demands into quantitative parameters, to deploy the functions forming quality, and to deploy methods for achieving the design quality into subsystems and component parts, and ultimately to specific elements of the manufacturing process.[17] While QFD was introduced in the 1960s, it required a fair amount of data to be collected to effectively analyze the functions and translate into design changes. The usage data, if properly collected and collated, can facilitate a much more rigorous use of QFD and related methods to improve engineering design. However, the engineering organization must install sensors in the product to collect the usage feedback, which collect data at the subsystem level to isolate system installation problems. If the security system installation of monitoring cameras was taking one full day, there must be a mechanism to study the time taken for each step, and to collect data from and these sensors to isolate bottlenecks in the installation process. It is often hard to collect data manually for feedback. How often do you get frustrated with navigation for a website, but decline to take the user survey? However, Cognitive Things are in the best position to tirelessly create the data needed for future analysis.

Another important aspect of engineering is the ability to experiment with the design and try multiple approaches, whereby field trials can be used to try many designs and to choose one that provides the best result. In a closed loop environment, such experimental designs can be carried out in the field and the feedback can be collected and analyzed for design optimization. Complex field tests with multiple alternative designs can be tested across an entire population of users, instead of small samples, as is often done today. Many software companies already carry out these experiments using beta testing programs, which give them the flexibility to offer new design ideas and to compare before making them a part of the general release.

Let us look at an example to illustrate cognitive device engineering. Network technologies for communications networks are evolving into cloud-based network function deployment, where each network function, such as routers and switches can be deployed via interchangeable cloud-based hardware that can be configured to provide specific functions. As smartphones and other

---

[17] Akao, Yoji (1994). "The Customer Driven Approach to Quality Planning and Deployment." Minato, Tokyo 107 Japan: Asian Productivity Organization. p. 339. *ISBN 92-833-1121-3.*

wireless devices (including Cognitive Things as wireless devices) proliferate and dominate network usage, the requirements for network resources may vary significantly with the users. If a group of 40,000 fans in a stadium start heavily using video replays in an important game, the network may be able to collect the usage information about the stadium, Subsequently, the network can be reconfigured to provide additional video bandwidth to the fans. Here the usage information may need to be collected, aggregated, and isolated to spot fans in a stadium. The information can then be channeled to the network engineering function, which can make on-the-fly design decisions to reconfigure the network.

Imagine household equipment that can learn from your past usage, and provide you with a button labeled, "same as before." Your car probably already has a navigation button for "Home" and can direct you home, or for that matter to many of your frequently visited destinations. In each case the Cognitive Things are being designed to learn and remember from past usage, and to make it easier to operate based on past usage.

2.5 Blockchain for Transaction Management

If things start transactions, how would they trust each other? The traditional model of contract management involves trusted third parties who witness or participate in the transaction. For example, in a transaction using a credit card, the buyer authorizes the seller to get paid using the credit card operator as the intermediary. The credit card operator performs this function by establishing a trusted relationship with buyers and sellers and maintaining a network with enough buyers and sellers to make such transactions viable. Globally trusted third parties conduct a large number of transactions, and assume the associated liabilities. In an asset sale, for example in buying and selling real estate in USA, title and escrow companies witness the transactions, and use notaries to verify signatures of buyers and sellers. The basic premise in a trusted transaction is the presence of this third party, who can verify the transaction and can participate in a dispute challenging the validity of the transaction.

As Cognitive Things get involved in transactions, these trusted parties are multiplying, often challenging human ability to comprehend and manage ledgers and transactions. When you use an Uber taxi to get a ride with a driver, the transaction involves your credit card company as well as Uber, both of them acting as intermediaries between the driver and you, each charging a percentage of the transaction and providing their service for a secured trusted relationship between two strangers. If you challenge the charges made by the driver, you can engage the driver, Uber and your credit card operator, each of them vouching for the validity of the transaction, which involved a series of digital events, witnessed and recorded by the trusted third parties. These events can be replayed in case of a dispute to check the validity of the charges.

Imagine that you're walking down a crowded city street, and a piano falls from the sky. As dozens of people turn to watch, the piano crashes down right in the middle of the street.

Then, without a second to lose, every person who witnessed the event is strapped to a lie detector and recounts exactly what they saw. They all tell precisely the same story, down to the letter. Is there any doubt that the piano fell from the sky? This is the principle behind the blockchain, a powerful invention that could profoundly affect our relationship with the digital world. A blockchain is essentially just a distributed ledger of digital events – one that's "distributed," or shared between many different parties. It can only be updated by consensus of a majority of the participants in the system. And, once entered, information can never be erased.[18]

A blockchain provides a secured way of managing transactions among strangers, with or without trusted third parties. In many ways, the blockchain is challenging today's financial institutions and governments by creating a distributed ledger, which can allow strangers to execute transactions. For machine-to-machine transactions backed by digital events, the blockchain provides a much-needed backbone for digital contract management and transactions.[19]

2.6 Chapter Summary

In this chapter, I have described cognitive devices, how they operate and how they are designed and engineered. So, let's apply the idea to imagine a Cognitive Television. The day I purchase and bring the television home, it can be plugged into the home entertainment system with an installation app on my smartphone. The smartphones of each household member can now become universal remotes for the television, remembering the programming and learning from the explorations. Each time I sit down to watch a program, it organizes the programming based on my past usage and also recommends new programming similar in content to the ones I liked previously. The content can be searched using unstructured conversations, where I can browse or

[18] Treasury Today, "Blockchain: explaining the future", Treasurytoday.com, September 2015, http://treasurytoday.com/2015/09/blockchain-explaining-the-future-ttti

[19] DeveloperWorks TV, "Blockchain Car Lease Demo", Mar 1 2016, https://youtu.be/IgNfoQQ5Reg

search based on my recollection of actors, scenes or storylines. The television may also recommend content based on what is popular to my micro-segment, or based on my viewing on other devices. The cognitive television senses how many people are in front of it and appropriately brings content based on the aggregate preference of the audience, differentiating between family viewing, romantic viewing and solo searching, for the new experience. The television is able to troubleshoot its problems, and optimize its experience based on bandwidth and number of televisions competing for the bandwidth.

Some of these functions are already available. Despite initial public backlash, Amazon, Netflix, and others prioritize contents based on history and recommend new content based on past viewing. As we proceed through the cognitive journey, we hope to see truly smart TVs that can operate, install, diagnose, and engineer themselves, optimizing the viewing experience. They will also collect feedback for better design of cable network, television or the universal remote.

The bigger prize is in using cognitive device log data to create smarter product engineering. Products often carry features that are sparingly used. Any installation or usage issues remain unacknowledged, and hence survive release after release. Usage data collection, analytics and field testing can significantly streamline product engineering. As products become digital and software controlled, product features and configurations can be dynamically reprogrammed. In this chapter, I used the example of telecommunications network where dynamic configurations can be used by the engineering organization to optimize the design for specific usage at a point in time.

As these 25 billion Cognitive Things provide many services to their 8 billion masters, what can they learn about their human masters, and how will they make their day-to-day tasks different? In the next chapter, I will explore the additional use cases arising out of man–machine interface as they impact individuals, and in Chapter 4, I will explore use cases associated with organizations.

3

Cognitive Devices as Human Assistants

3.1 Introduction

Race Across America (RAAM) is the longest running endurance bicycling race in the world.[1] It covers 3000 miles coast to coast, and requires an ability to compete almost non-stop, with minimal rest breaks. It requires the cyclist to cover 12 states, including the Pacific Ocean coastline, Rocky Mountains, Great Plains, and the Atlantic Ocean coastline. No wonder 300+ racers from 25 countries participate with the help of 100+ crew. Let me introduce Dave Hasse, who has successfully competed in this tough race multiple times. Dave Hasse's love for biking developed late. Dave found ultracycling after competing in a 24-hour race on a whim. When he found out it was a RAAM (Race Across America) qualifier, his reaction was, "If you qualify, I guess you have to do it." An attitude that sums up what Dave is all about. After nearly riding himself to death in the 2004 Race Across America, and doing so on camera for all the world to see in Steven Auerbach's documentary, Race Across America, Dave went on to finish the RAAM three times over the next five years.[2]

David Haase has always relied on gut feeling when making critical decisions during races. However, for his 2015 attempt to become the first American to win the extremely competitive Race Across America, Dave turned to IBM to help him make crucial decisions in real time. By combining his own intuition with insights from high-speed, high-performance analytics, Dave was

[1] RAAMmedia, "2015 RAAM Sizzle Reel", YouTube, Sep 22, 2014, https://youtu.be/s-AGsrtll6w

[2] From Dave Hasse's website. http://www.davehaase.com/

A. Sathi, *Cognitive (Internet of) Things*,
DOI 10.1057/978-1-137-59466-2_3

able to adapt to evolving race variables and perform at a more advanced level than ever before. These Cognitive Things supported Dave by monitoring his physiology, as well as environment changes, including wind changes and temperature prediction. As he was racing about 22 hours each day over eight days, twenty hours and six minutes, his downtime was optimized via predictive algorithms, which took into account his fatigue, terrain, and weather predictions.[3]

In this chapter, we will explore how Cognitive Things can assist individuals to excel in their chosen pursuits, by augmenting their intuition and decision-making. David's situation is unique, but it expresses how much an individual can push the boundary of physical limits in a competitive situation. However, he needs to make a number of critical decisions—such as how hard to pedal and when to take a break, which will help him optimize his performance. In last year's race, there was a windstorm, posing a serious headwind problem for the racers. Contrary to conventional intuition, the winners rode their bicycles for much longer in their first leg, leading to a significant headwind difference. As technology becomes available ubiquitously, everyone has a digital device, which can provide access to large amount of data, like the weather and terrain information. Access to the data is the first level of support. Cognitive devices help humans decide on alternatives that can significantly improve outcomes through better utilization of available data. They do so by analyzing and correlating across disparate information sources, scoring alternatives in real time and designing actions.

Eric Horvitz, a technical fellow at Microsoft, presented a model of collaboration in his TED talk show, where human cognition is supported by machine intelligence in three areas: decision, attention, and memory. The human mind is extremely good at making connections, processing information, and making decisions. Machines provide the much-needed support by laying the foundation for systematic decision-making through careful evaluation of alternatives, wading through large volumes of data and providing attention to the most significant information, and memorizing and recalling a lot more information as a loyal assistant to the human they support.[4]

It is important that a machine is loyal to the human they serve. In the process of receiving "free" downloads of apps, you may be falling a victim to a disloyal machine, which trades your data with someone else to provide you with either poor or unneeded advice. For wide acceptance, the device must

---

[3] IBM Analytics, "Insight trumps luck in the Race Across America", YouTube, Sep 2, 2015, https://youtu.be/TWU8QzJi3TM

[4] Eric Horvitz, "Making Friends with Artificial Intelligence", TEDx Talks, Feb 19 2013, https://youtu.be/dpoVh9xwdD4

be easy to use, trustworthy and capable of working at the appropriate pace. I will describe a couple of good and bad examples of experiences that can be received by humans.

3.2 Shopping and Buying Assistant

Shopping for cars was a difficult experience for me. Dealers cherished the information they carried about MSRP, dealer margins and financing options. At each step of the bargaining, the car salesperson pushed hard to get a commitment, and I could not collect enough information to evaluate alternatives and make the optimal purchase. Prior to the availability of Internet, I used to gather a series of experts around me to help me bargain to get the best deal. Information was exchanged in a guarded way and carefully used to get the best deal.

As information becomes ubiquitous, this area is under intense experimentation and the tide is finally turning towards the buyer. A number of buying assistant apps has leveled the information, giving me a perfect opportunity to fully understand the inventory availabilities, manufacturer offers, and negotiation margin. However, there is one piece of information, I would rather not tweet—my intention to purchase the car along with my cell number. The resulting calls I am likely to receive will overwhelm my curious state of mind, especially if I change my mind or purchase a car. An app promising me buying assistance did a rough equivalent of this ghastly act. When I used this app to get advice, I was shocked to find my intention to purchase a car was broadcast to all the local dealers, and I was flooded with dealer calls, even a week after I had purchased my dream car. As I delved into the detail, I found the app was challenged by its business model, and its owners had decided to side with the dealers to trade consumer data for short-term profits. Unfortunately for them, they received a poor rating from me and social media was also buzzing with negative comments. Is there a buying assistant that is discreet, informative, and a fair representative of the buyer's view? As buying assistants mature, they are finding that consumers will share information only if they trust their assistant. Are there any trusted buying assistants?

Fortunately, these trusted agents are rapidly offering enormous support to price-sensitive shoppers. If you've got your eye on something you want to buy, sometimes it pays to hold off. Other times, your best bet is to make the deal as soon as possible. How do you know when to pull the trigger? One tool that can help is the price tracker. Sites like Camelcamelcamel.com and PriceZombie.

com can help you gauge whether to buy something now or later by showing you month-by-month price variations and calling out high and low points during the year. That way, you'll know a good price when you see one. Both sites also will send you price-drop alerts. CamelCamelCamel shows you charts of Amazon prices only; PriceZombie covers dozens of retailers. You can also use tools like Pinterest's alerts and the Slice app to get notified about price drops.[5]

During a pre-holiday shopping trip to New York, Lisa Libretto received an enticing alert on her iPhone: an offer for a $25 discount on a Vince Camuto handbag that she had coveted on the retailer's website. "If my phone is alerting me to the discount or some information about items I might like, that will totally pull me in," said Ms. Libretto, who lives in Ridgefield, Connecticut. The alert arrived at an opportune time, pinging as she neared the entrance of the Vince Camuto store. And it cemented her decision: she would buy the purse after all. But the timing was no coincidence. The app that she had downloaded from ShopAdvisor used beacon technology, a new addition to location-based marketing, to pinpoint her whereabouts before sending the discount. "I hope it's a technology more companies will use," she said.[6]

ShopAdvisor, a four-year-old company based in Concord, Massachusetts, has added a wrinkle to location-based mobile marketing that it hopes will be the breakthrough retailers are seeking. GPS-based mobile apps are not new and geo-fencing, the ability to create a virtual perimeter around a designated location such as a shopping mall, has given retailers the ability to send push alerts to prospective customers nearby. But beacon technology can pinpoint a customer's location so precisely that a retailer knows when that shopper is lingering in the shoe department or browsing in lingerie. ShopAdvisor, which offers its own mobile shopping app and specializes in creating multichannel mobile shopping platforms for media companies as well as retail brands, has incorporated beacon technology in a novel way.

Cognitive selling is gradually becoming mainstream. For example, Taco Bell is using a cognitive application to take care of routine ordering. Taco Bell unveiled the TacoBot, a Siri-like version of the cashiers that take your order at its restaurants. Taco Bell built the bot for workplaces that use Slack's messaging platform to communicate internally. Now, instead of someone jotting

[5] Lisa Lee Freeman, "5 Money-Saving Tools to Try Before You Buy Anything", Money, December 15, 2015, http://time.com/money/4147801/money-saving-sites-apps/

[6] Glenn Rifkin, "Enter the Shoe Aisle, Feel Your Phone Buzz With a Personal Deal", The New York Times, December 30, 2015, http://www.nytimes.com/2015/12/31/business/smallbusiness/shopadvisor-lets-retailers-target-shoppers-by-location-and-interests.html

everyone's orders on a Post-it and hoping the drive-thru attendant gets everything right, they can ask TacoBot to put in the order for them. "We are at a point of switching up how we use computers," said Martin Legowiecki, senior VP and creative technology director of Deutsch, the agency that built TacoBot, "It used to be we had to talk like computers." Now computers are increasingly talking like us. TacoBot can take down your customized order and even crack a couple of jokes. For example, if you tell TacoBot that you're drunk, it'll add a cup of water to your order. And if you ask for a food recommendation, it'll give you one in binary code that translates to Doritos Locos Taco.[7] The Deutsch team built TacoBot using Wit.ai, an online service for building software that's able understand and respond to human language. (Facebook bought Wit.ai to help build its virtual assistant M). Legowiecki says this means it will be able to bring TacoBot to other apps in the future, such as Hipchat, Facebook Messenger, Amazon Echo, and even Apple TV. Facebook has unveiled a Messenger platform for building bots that can interact with customers or handle purchases.[8]

3.3 Care Assistant

During the twentieth century, the number of persons in the United States under age 65 has tripled. At the same time, the number aged 65 or over has jumped by a factor of 11! Consequently, the elderly, who comprised only 1 in every 25 Americans (3.1 million) in 1900, made up 1 in 8 (33.2 million) in 1994.[9] Those older US residents are expected grow from 43 million in 2012 to nearly 84 million over the next four decades as the baby boomer generation ages. One in 5 of the nation's population will be 65 or older by 2030, the year by which all baby boomers—so-called due to the "boom" in US births in the years following the Second World War—hit the unofficial retirement age.[10] Many of these elderlies are living alone or with a significant other and rely on their support system for medical or other needs. My dad lives with me and continues to show a remarkable long-term memory at the age of

[7] Klint Finley, "The Future of Work Is ... Ordering Taco Bell Through Slack?", Wired, April 7, 2016, http://www.wired.com/2016/04/future-work-ordering-taco-bell-slack/

[8] Danny Sullivan, "Facebook's Messenger Platform launched, allows businesses to build bots", Marketing Land, April 12, 2016, http://marketingland.com/faceboook-f8-172675

[9] Economics and Statistics Administration, "Sixty-five Plus in the United States", US Department of Commerce, May 1995, http://www.census.gov/population/socdemo/statbriefs/agebrief.html

[10] Susan Heavy, "Number of U.S. Elderly to double by 2050—Reports", Reuters, May 6 2014, http://www.reuters.com/article/us-usa-aging-census-idUSKBN0DM1BS20140506

93, remembering minute details from events that happened decades ago. The recent past is a different story. Over the last couple of months, he constantly needs assistance in remembering daily tasks like taking medication twice a day, or regular physical activities with a target. How can Cognitive Things be of assistance to these elderlies and lend them some assistance?

The house could be set up with a set of sensors that track activities, such as the opening and closing of refrigerators, microwave, and cabinets. As the elderly person conducts his daily tasks, the sensors keep track of those activities by time of day, thereby creating a log of daily activities. By aggregating events captured by these sensors, the elderly care system can infer whether the elderly person ate his food or took his medicine, performed physical activities, or washed his hands after visiting the toilet. This information can be compared across a social group and reported to care givers.

AT&T is offering a Digital Life solution for elderly care. Digital Life Care is designed for "informal caregivers," in other words, children who want to keep an eye on their relatives' health without moving in with them or shunting them off to assisted living.[11] It won't actually take care of them, but it will alert the caregivers as to whether or not the elderly folks are taking care of themselves.[12]

Let us now consider another care situation that requires some careful attention and solution development—emergency care. The number of 911 calls placed by people using wireless phones has significantly increased in recent years. It is estimated that about 70 percent of 911 calls are placed from wireless phones, and that percentage is growing. For many Americans, the ability to call 911 for help in an emergency is one of the main reasons they own a wireless phone. However, the mobility of wireless telephone service makes determining a wireless caller's location more complicated than determining a traditional landline caller's location, where numbers are associated with a fixed address. In order to enhance the ability of emergency personnel to respond efficiently and effectively to callers placing wireless 911 calls, the US Federal Communications Commission (FCC) has taken a number of steps to ensure that wireless service providers make location information automatically available to Public Safety Answering Points (PSAPs).[13] These rules will be phased in over time. Unfortunately, at the time of writing this book, the wireless

---

[11] AT&T, "Connecting Caregivers with AT&T Digital Life Care", YouTube, Sep 8, 2014, https://youtu.be/2AUIMkhszl8

[12] Sascha Segan, "AT&T Will Keep Your Grandma Out of the Nursing Home", PC Magazine, September 9, 2014, http://www.pcmag.com/article2/0,2817,2468204,00.asp

[13] "Emergency Communications", on the Federal Communications Commission site, https://www.fcc.gov/consumers/guides/911-wireless-services

location remains approximate. There are many sources of wireless location, each with its deficiencies. The PSAP is provided with the location of the nearest wireless antenna, which could be half a kilometer away. If the phone is GPS enabled, and PSAP can access GPS location, that would provide more precise latitude/longitude coordinates. Unfortunately, GPS does not provide a good estimate of Z coordinate or vertical height, which would be important in the case of multistorey buildings. Also, GPS is a battery devourer. For indoor phones connected via Wi-Fi, their location can be pinpointed using a Wi-Fi transmitter location.

For global travelers, emergency care does not have one universal number. Each country has its unique number and process for emergency care. If I am in need of emergency care, I probably do not have the time to find the correct local number.

Fortunately, the need for help in emergency situations is a good reason to share location information. Also, location can be inferred with analytics and predicted with a reasonable accuracy for most consumers. An ideal emergency care system would optimize location prediction using multiple sources, using historical location data as a predictor, and may also quiz the smartphone for additional information. It would provide a cognitive user interface to the consumer seeking emergency care, as well as to PSAP and emergency care vehicles trying to reach the distressed person(s). The system would work seamlessly across national boundaries with many different emergency systems.

3.4 Travel and Entertainment Assistant

The company, Uber, is less than five years old and already popular in the mainstream for providing travel assistance and for connecting a traveler with a Uber driver providing taxi services. The concept is rapidly being copied by the mainstream taxi services, which are offering the ability to locate a taxi driver while providing their own trusted brand for the travel experience. In either case, it is an interesting application of location and contact information sharing. The Uber driver gets to know the location of a potential traveler and vice versa. The traveler can call the Uber driver by using the phone number provided on the app. In my first experience, I felt uncomfortable sharing location and contact information. In fact, in one taxi app from a source other than Uber, I did get unsolicited text messages from the taxi drivers a month later. However, Uber seems to have created a layer of masking to make it harder for the Uber driver to invade my privacy. Uber uses a call transfer to obfuscate the phone numbers. The number I see for the Uber driver is not their real number, and they do not get to see the traveler's number either. Depending on the local taxi regulations, Uber may need to share the first and last name of the traveler with the driver.[14] Uber is working with local regulators to remove such rules. So, in this case, the service is easy to use, and is doing its best to win trust, despite strong lobbying from taxi organizations.

Let me now move on to assistance while staying at a hotel. I travel nearly once a week and when the travel is within USA, it lands me at a hotel room typically around midnight given that my starting point is California and I am traveling against the sun. However, I do like to visit the fitness center for a nice midnight jog or walk. What is the likelihood that I will find the fitness center by just following directions posted on the hotel walls? Then, after the jog, I need a bottle of water, and sometimes do not have any Wi-Fi directions, or fresh towels. You get the picture. Each of these questions is a repeated call to the front desk, often when the front desk is also dealing with other late-arriving guests, and the wait reduces my cherished sleep time. In the last IBM Insight conference in Las Vegas, I was delighted to hear Raj Singh, the Chief Executive of Go Moment, in the main tent as he demonstrated Ivy, a cognitive assistant, which can be called upon to help. Creating Ivy has required its creators to get down and dirty with the individual issues that come up at each particular client hotel. In doing so, they've found that there are about 1,400 issues that can crop up, barring an outlier event such as a fire. According to Singh, more than half of these don't require human judgment on the other side, and that half account for a significant majority of the interactions that come up. Ivy is powered by IBM's Watson, the world-famous AI engine. Ivy uses the Watson engine to decipher and fulfill a wide variety of requests from hotel guests in real time.[15]

Figure 3.1 shows Ivy in action in respect of two customers. The interaction is via messages. Hotel staff may be asked to join in, if the discussion requires their assistance, while Ivy handles the most repetitive questions.

As a hotel guest, I get instantaneous response to my questions. Difficult questions are deflected to the hotel staff. It certainly will get hotel brownie points for being responsive and will reduce negative publicity in social media. Equally important, the hotel staff are now relieved from repeated calls on subjects that can be handled automatically, giving them the time to deal with

[14] Polly Mosendz, "Uber drivers have privacy problems too", Newsweek, Nov 19, 2014, http://www.newsweek.com/uber-taxi-e-hailing-riding-app-travis-kalanick-emil-michael-josh-mohrer-uber-285642

[15] Micah Solomon, "How IBM Watson Is Poised To Transform Customer Service And Hospitality Via AI", Forbes, December 17, 2015, http://www.forbes.com/sites/micahsolomon/2015/12/17/ibms-watson-ai-is-poised-to-transform-customer-service-and-hospitality/#2fb291361b13

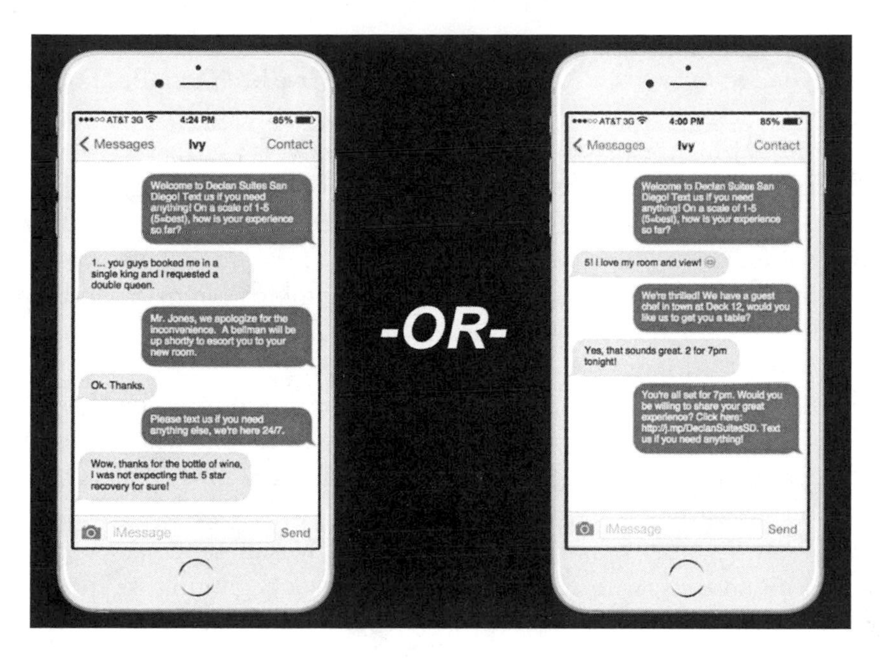

Fig. 3.1 Ivy in action

the more complicated issues and improving their productivity in dealing with guests on a busy day.

3.5 Administrative Assistant

Most people have more manual tasks than they would like. Can a personal administrative assistant help automate these tasks? For example, I would like to know where I could eat nearby. I can possibly go to Yelp, read a bunch of reviews, look for directions and find something I like, or just give a command to a personal assistant to do the same. Something as simple as finding the current local weather may require a couple of clicks through a weather app, or a voice command to the personal assistant can achieve the same. Could this personal assistant act more like a human and respond to the variety of requests I may be making? As it turns out, human administrative assistants are very good at dealing with a variety of such requests and it is not easy to build an automated agent that can replicate the same level of dexterity. There are many applications vying to be your personal assistant—for example, Siri, Google Now, Cortana, and Alexa. However, providing a quality cognitive personal assistant is not easy.

Facebook has begun to test its new virtual personal assistant, M, in a small beta beyond its own walls. And the early reports have been ecstatic. Tales have emerged of tickets booked and food ordered. M, we are told, is already amazing at minimizing the pain of certain small tasks. M is a virtual assistant, but it's not pure artificial intelligence like Siri. Facebook has cleverly positioned M as part AI, part human. This invites you to let your imagination run wild about the amount of work being done by robots.[16] Facebook is attempting an interesting way to build its assistant. It uses a combination of AI and a person behind the screen (much like the wizard in *The Wizard of Oz*) in case the automated agent cannot provide a proper response.

So far, these assistants have been dealing with adults. John Anderson is a cognitive psychology professor at Carnegie Mellon who built some of the early cognitive language learning models based on how children learn to speak. My first exposure to cognitive psychology was through his course, and his story telling was captivating. He had good samples available, as his son Jay, who was born in 1980, was beginning to speak and John Anderson decided to use his ACT cognitive model to replicate Jay's learning. Jay was learning a lot, and at the age of two he surpassed the total disk space available to John, which in those days was measured in megabytes. Interestingly, Jay's first two words were "more gish", which translated meant "I want more cheese". Utterances generated by children are one word for a long while, and then two words for quite a while; only after considerable experience do long utterances begin to appear.[17] Even when they start speaking complex sentences, children often deviate from normal in their pronunciation. A cognitive assistant for a child must deal with children's learning cycles, which differ considerably from adult learning. John Coolidge and JP Benini launched a toy dinosaur toy driven by IBM's Watson software.[18] Developed under the aegis of a company called Elemental Path and a project called Cognitoys, this tiny plastic dinosaur uses speech recognition techniques to carry on conversations with kids, and according to Coolidge and Benini, it even develops a kind of smart

---

[16] Nathan McAlone, "Facebook's new virtual assistant 'M' isn't actually a robot—does that take the fun out of it?", Business Insider, Oct 28, 2015, http://www.businessinsider.com/facebook-m-virtual-assistant-is-mixture-of-ai-and-human-answers-2015-10

[17] John R Anderson, "The Architecture of Cognition", Harvard University Press, Cambridge MA, 1983, http://www.amazon.com/Architecture-Cognition-John-R-Anderson/dp/080582233X/

[18] Elemental Path, "Elemental Path presents CogniToys", Vimeo, Feb 24, 2015, https://vimeo.com/120550878

personality based on likes and dislikes listed by each child.[19] I will describe Cognitoys in more detail in Chapter 2.

3.6 Chapter Summary

Individuals are gradually getting immersed in a tsunami of data from their devices. Often, there is too much data and not enough information. How about using Cognitive Things to derive information from this data, analyze the information, and use it for making recommendations? The decision rests with the individual, but it is based on a thorough and detailed analysis of the facts rounded by an untiring cognitive machine. I started this chapter with the amazing story of David Hasse. He is the one who decided how to ride his bicycle across the continent, with tremendous determination and super-human endurance. Cognitive Things provided a small yet important contribution through organizing his terrain, weather and movement information by giving him interpretations, analysis and recommendations instead of raw data. In his limited rest time, David was able to focus on the race and the strategy and not on tactical and administrative tasks.

In this chapter, I have described how the cognitive devices support humans, including intelligent care assistants, shopping assistants, travel and entertainment assistants, and administrative assistants. Each of these assistants is specialized to certain task areas and is capable of providing assistance by consistently adding three cognitive skills to the user's tasks—the ability to observe and collect events, the ability to summarize and analyze these events, and the ability to share events and augment them with other information. All of these are very information-intensive activities, which can be performed by the most organized consumers, but are hard to be performed consistently and with good precision. The assistants make it convenient for the consumer to conduct their tasks without constantly organizing and collecting information.

In the next chapter, we will explore how organizations can make use of such assistance. Although the assistants perform very similar activities, the results are an increased and optimized performance for the organization across a diverse set of departments and use cases.

[19] Davey Alba, "A Toy Dinosaur Powered by IBM's Watson Supercomputer", Wired, Feb 17, 2015, http://www.wired.com/2015/02/cognitoys-ibm-watson/

4

Cognitive Things in an Organization

4.1 Introduction

What are the organizational functions supported by Cognitive Things?

How do Cognitive Things support Smarter Operations?

Could the Cognitive Things transform engineering function?

How do Cognitive Things support Contextual Marketing?

Could the Cogntive Things accelerate smarter care?

How do Cognitive Things detect and control theft and fraud?

Would it be possible for the communications department to be proactive, so the network coverage is fixed as the team gets built, local hotels are approached by the corporate travel department for hotel discounts and a concierge is set up for employees, who are invited to join the group based on their frequency of travel to the project site? All of this requires a travel organization to observe the project team growth, and proactively participate in setting up the support services. Mobile employees have special requirements, whether sales, fleets, or consultants.

A. Sathi, *Cognitive (Internet of) Things*,
DOI 10.1057/978-1-137-59466-2_4

I have been a mobile employee all my life. It was common for me to travel three to four times a week, each trip requiring multiple days of hotel stays. Travel departments often send survey questionnaires to ascertain issues faced by travelers like me. Even if I filled in a survey, by the time the decisions were made, I was in another part of the world, and no longer traveling to the location where they fixed the hotel availability. Wireless coverage is even harder. Unfortunately, wireless service provider's corporate campuses were my most frequent hangouts. Sometimes I was lucky to visit the wireless provider who was also providing service to my phone and I was delighted to see five bars in their corporate headquarters. Unfortunately, more than likely I was also visiting a number of their competitors, and had to be content with the occasional single bar on my phone at their office locations. Why would my wireless service provider locate antennas close to its competitors' offices? I often marveled at the large population of traveling consultants, all power users and heavy spenders and yet unloved by their carrier because they chose to visit their competitor! Could the telecommunications providers and the corporate telecommunications purchase departments indulge in proactive care for these mobile workers?

In the last chapter, I discussed how Cognitive Things are able to collect data about consumers and support collaborative decision-making. Can we apply the same principles to corporate Cognitive Things? What use cases and business drivers can be used to facilitate widespread deployment of Cognitive Things for corporate collaboration? Inter-department communication and collaboration is always much harder than intra-department communication. Information moves very slowly, even in post-start-up small organizations. It is difficult for large corporations to obtain quality information to make operational decisions. Marketing departments find it easier to hire external consultants to get data about their products than ask their own engineering staff. Even if the information were made available, its data collection, organization or governance may not suit the company, making it extremely hard, if not impossible to integrate across departments. Cognitive Things are bringing two important capabilities to transcend corporate walls. First, they are able to observe a lot more data about customers, products, employees and operations, which can be shared throughout the corporation. Second, they are able to transport insight and cognitive decision-making from one corporate division to another, bringing them closer to the data collection points for real-time context-specific decision-making.

Fortunately, in many industries the Internet of Things (IoT) is already being used to collect a fair amount of data. In some cases, the IoT is embedded in the product to its consumers; for example, automobiles, home appliances and smartphones, to name a few. In other cases, the IoT is part of an automation program, such as manufacturing automation, distribution, or digital engagement. In many of these cases, the data collection platforms, governance across divisions, or analytics activities, may already be available.

This chapter covers five major use cases for Cognitive Things. First, I will discuss smarter operations and how Cognitive Things can use their sensors to collect operational details, which can be used for improving the operations. This is a classic business process improvement use case, where the Cognitive Things provide elements of information collection as well as analytics and interpretation of data to offer major changes to the operations. The second use case concerns smarter engineering, which requires collecting data about product usage and component failures to design an improved product. In this use case, product data is reorganized for improved engineering. The third use case is related to contextual marketing. This use case brings customer information together with product usage information to make inferences regarding the suitability of a product for a micro-segment of customers. The fourth use case uses Cognitive Things to improve customer care. In the majority of post-sales operational support, troubleshooting or help desk functions, Cognitive Things can help automate care and enhance self-service by collecting information about how the product is actually used in the field and related challenges faced by the customer. Last but not the least, the fifth use case deals with the finance function and explores ways to use Cognitive Things to reduce revenue leaks and counter fraud management. These use cases are summarized in Table 4.1. As you can see from the table, there is a fair amount of overlap in the entities involved, and cognitive infrastructure can be easily extended to cover multiple use cases.

Table 4.1 Sample organizational use cases

Corporate function	Use cases	Entities involved
Operations	Process re-engineering/improvement, infrastructure management	Process flows, infrastructure
Engineering	Product improvement, component evaluation	Product usage, component performance
Marketing	Contextual promotions, advertising	Customer 360, needs/demand, intent to purchase, product usage, location
Care	Self help	Product usage, location, current purchases
Finance	Counter fraud management, revenue assurance	Customer 360, product usage, location, current purchases

4.2 Smarter Operation

What bugs people the most about grocery shopping? It's not the in-store Muzak or the occasional squished loaf of bread. It's the dreaded wait at the checkout line, according to Kroger customer surveys, prompting the supermarket chain to test a variety of technical solutions over the years.[1] If you live near a Kroger grocery store in the USA, you may have observed a big screen showing a prediction of the number of checkouts needed. In a typical queuing system, the average wait time can be derived from three parameters: average service time, average arrival time and number of servers engaged in providing service to the customer. At the grocery store, they can redeploy their clerks for check-out operation or for store operations. Ideally, they should minimize average wait time as well as the total number of clerks employed at the store. Cognitive Things provide the ability to optimize across these two variables.

Kroger thinks it finally has the right mix of technology: QueVision, which combines infrared sensors over store doors and cash registers, predictive analytics, and real-time data feeds from point-of-sale systems. From the moment customers walk through the door of a Kroger store, the QueVision technology works toward one goal: ensuring that they never have more than one person ahead of them in the checkout line. The technology, now deployed at more than 2300 Kroger stores across 31 states in the USA, has cut the average wait time from more than 4 minutes to less than 30 seconds, the company says.

There are observations, which make it easy for the grocery store to predict the number of customers arriving at the checkout counters. Grocery stores use beacons to count the number of customers entering and leaving the store. By analyzing historical data by time-of-day and day-of-week, they are able to predict the amount of time each customer is likely to spend in the store shopping for their groceries. Using this prediction model applied in real time to the number of customers entering the store, the grocery store predicts the number of checkout registers needed in the near future. This information is displayed on the television screens near checkout counters, giving the employees an early warning on the number of checkout registers needed before the customers arrive and start queuing. Thus, checkout registers are only open

[1] Laurianne McLaughlin, "Kroger Solves Top Customer Issue: Long Lines", Information Week, April 2 2014, http://www.informationweek.com/strategic-cio/executive-insights-and-innovation/kroger-solves-top-customer-issue-long-lines/d/d-id/1141541

when the customers start lining up, and the employees can use the rest of their time to focus on other operational tasks with minimal idle cash registers waiting for customers.

Beacons are getting popular as crowd management devices and have also been used effectively for informing drivers about the number of open parking spots in a given floor of a parking lot. However, beacons have their limits. They cannot differentiate across different types of customers and do not form a relationship with their customers. In a smart stadium, there are different types of customers coming to watch an event, each with a pre-specified seat and a set of gates from which they can enter the stadium and get to their seats. If the stadium establishes a personal relationship with the guest, this relationship can be used to upsell merchandise, provide information regarding food stalls, or provide event replays later. The cell phones connected to the stadium Wi-Fi spot can be used as smart beacons to build this one-to-one relationship and can support the guest with enhanced engagement at the stadium. As the guest approaches the stadium and enters the Wi-Fi area, they can be provided with an option to use the stadium Wi-Fi supplying information on how best to get to their seats with minimal queues. From here onwards, the Wi-Fi system can provide accurate information to the stadium regarding each guest who has opted for the Wi-Fi, and can use this information, along with the stadium's app to provide personalized experience to each guest.

The Sun Life Stadium, home of the Miami Dolphins, is an unusual location in South Florida that spends all but 65 hours a year empty. But during that brief window of time, it's a bustling hub of competition, celebration and consumption. About 10 times a year, some 70,000 of its citizens make the pilgrimage from the surrounding region onto a half-a-square- mile plot that many consider sacred. There was a time when selling more beer and hot dogs might have been good enough revenue enhancers. But today it's a far more complex equation, with teams now obliged to think about their fans' total experience. After 17 years of playing quarterback for the Miami Dolphins, NFL Hall of Fame inductee Dan Marino knows something about keeping Dolphins fans loyal. While winning helps, Marino explains, true loyalty comes from the quality of the game experience, and from the sense that the team cares about delivering that experience. "On game day, it starts outside the stadium—in the traffic flow, the parking, getting through the gate quickly, finding their seats and having their favorite food and beverage close by. All those factors help to build loyalty," says Marino. Indeed, between the trips to and from the game, the fans' experience within the stadium is also marked by mobility—getting from the parking lot and then through the turnstiles to one's seat, and then on to the restroom, to concession stands, then back to the car at the end of the game.[2] And like anything in motion, friction points like long lines can disrupt and slow this traffic flow, with adverse impacts on fan satisfaction and game-day revenue.[3]

Wireless networks and cell phones can be used for mobility data about people or Cognitive Things. Location data from cell towers can be used to observe and analyze the mobility of these cell phones. Smart cities are beginning to use mobility data in innovative ways to study and analyze their road infrastructure. They have been using many traditional ways of observing road conditions. Cell phones offer innovative ways of collecting far more observations at a much better economy of scale, as illustrated in the following case study from Nairobi.

Africa is IBM's 12th global research lab and the first industrial research center in Africa. With facilities in Kenya and South Africa, it is developing commercially viable solutions to transform lives and spark new business opportunities in areas such as transportation, water and energy. Nairobi is one of the world's fastest-growing cities, which places increasing pressure on its systems and infrastructure. Experts estimate that Nairobi loses up to US$ 1 million every day due to the congestion's impact on lost productivity, fuel consumption, traffic accidents, air quality, and many other issues.

Dr. Aisha Walcott Bryant, Researcher at IBM, says: "Conventional traffic management relies on citywide sensor networks, which are not feasible in cities where the road maintenance budget is tight." Such solutions also fail to take account of common factors that influence traffic in many developing cities. For example, heavy goods are often transported on low-grade roads, creating potholes that can turn major streets into single-lane passages. Bad weather can make whole sections of road impassable for days at a time, completely changing the way traffic flows. "And in many places, speed bumps have been installed—in some cases by unauthorized parties, in haste, and not up to standard. Potholes and poor quality, unmarked speed bumps can cause damage to vehicles as well as impacting traffic flows, and there is no central registry of where they are located." Road conditions have a wide range of impacts on the lives of Nairobi's citizens, the efficiency of its businesses and the ability of its government to drive change. Where road conditions are poor or unknown, it is hard to manage traffic effectively. This leads to congestion and dangerous driver behavior, which

[2] IBM Client Voices, "IBM & Flagship Solutions Partner with Sun Life Stadium to improve efficiency and fan experience", YouTube, Dec 13, 2012, https://youtu.be/1dsIYJfyR5I

[3] Smarter Planet Leadership Series, "Sun Life Stadium: Learning from cities to improve game day for fans", IBM, February 2013, http://www.ibm.com/smarterplanet/global/files/us__en_us__leadership__sunlife_stadium.pdf

delay vital civic services such as garbage collection—and in turn, raise public health issues and impact business.[4]

Nairobi City County recently invested over US$ 3 million in new waste collection vehicles—but due to the difficult road conditions and traffic, the new trucks were struggling to deliver the improvements in waste management efficiency that the city had expected. Realizing that IBM Research was already working on the city's traffic problems, the head of the Department for Environment, Water, Energy, Forestry and Natural resources met with the IBM team and asked them to help find a solution. Dr. Walcott Bryant comments: "When the city asked IBM to help improve the efficiency of its garbage trucks, we realized that the trucks themselves could help us solve the traffic problems by collecting reliable data about road conditions across the city."

IBM Research Africa's mobility team realized that a locally relevant solution was the key: Nairobi needed a method of capturing data about road and traffic conditions that would not require the investment and maintenance costs of a traditional solution.

The team came up with an elegant, low-cost answer: they installed adapted smartphones in a number of the city's waste collection trucks. As the trucks drive around the city, the phones' sensors (such as accelerometers, magnetometers, gyroscopes and GPS) stream data about each vehicle's location, speed, acceleration, and vibration levels to an IBM analytics solution in the cloud. Dr. Walcott Bryant explains:

We are creating a digital map of the city, showing hazards that cause drivers to brake or swerve such as speed bumps and potholes.

We can also see how long the trucks are spending collecting refuse, in traffic, at the dump and off-route. This has transformed our understanding of how road conditions, congestion, waste management, and a host of other factors interact to create Nairobi's complex traffic situation. More importantly, it has already given us a lot of insights into how that situation can be improved.

We identified a bottleneck in the waste disposal process: garbage trucks were spending between two and four hours per day at the dump. Most trucks were only able to visit the dump once per day, due to its location, traffic congestion, and sometimes because the trucks were going off-route to undertake other jobs.

Drivers also typically spent up to two hours per day in traffic, wasting time and fuel. One of our goals is to optimize routes according to traffic levels and instill

[4] IBM Research, "Living Roads", YouTube, Jul 16, 2015, https://youtu.be/09S5J-asEXs

transparency into the fleet management process. This will help the city complete more collections using its existing fleet—and therefore keep Nairobi cleaner and safer for its citizens.

The hazard map also enables authorities to improve road maintenance planning. For example, in one sub-county alone, the map has highlighted 750 potholes, which had been causing damage and delays to garbage trucks and other traffic. "We're in discussions with city authorities and private companies about how the solution could be used to optimize routes for emergency services and delivery vehicles," says Dr. Walcott Bryant. "It's opening up so many opportunities—not just for Nairobi, but for other emerging cities around the world." The solution offers huge potential for Nairobi to become an exemplar African smart city by using a cost-effective and smart approach for gathering actionable insight about traffic flows, while also enabling smarter decisions about street planning and road safety. By building fleet services that capture real-time traffic dynamics, road surface conditions, and driver behavior at low cost, IBM and Nairobi are demonstrating to the world how locally relevant, contextualized innovations can increase government efficiency, provide cost-savings across sectors, and create new potential revenue streams.

Let me now take you into a case study about commercial fleets.

International automotive supplier Continental AG is showcasing the forward-looking capabilities of its eHorizon platform. Thanks to this software, fleet operators using Scania trucks have already saved over 63 million liters of diesel (equivalent to €86 million) since 2012 according to an estimate from Continental. "In principle, the eHorizon uses map data to give vehicle electronics a glimpse into the future. This allows the vehicle to adjust to the upcoming route early on and actively reduce consumption," explains Helmut Matschi, member of the Continental Executive Board and head of the Interior division.[5]

Driver assistance systems or actuator behavior such as braking and steering can be prepared for upcoming traffic situations, long before the vehicle sensors detect the situation. It takes into account dynamic events such as weather, accidents, or traffic jams. Knowing what is ahead on the road allows drivers/vehicles to adjust their driving style to be safer, more economic, and more comfortable. Continental's eHorizon utilizes digital maps to anticipate the road ahead. It utilizes real-time digital maps to expand the digital horizon. The maps are enhanced using self-learning through data collected during

[5] Continental Corporation press release, "CES 2015: Dynamic eHorizon from Continental Points to the Future", Dec 10, 2014, http://www.continental-corporation.com/www/pressportal_com_en/themes/press_releases/3_automotive_group/interior/press_releases/pr_ehorizon_ces_2015_en.html

the drive and real-time correction of map databases. Additionally, it extends and enhances the navigation with dynamic information from external sources such as road infrastructure or crowdsourced from vehicles.

These maps are provided by HERE,[6] a provider of precision 3D maps. Precise maps, combined with in-car connectivity, create endless possibilities. For example, with an exact picture of the road, a hybrid car will be able to more efficiently switch between electricity and gas. Car manufacturers will be able to reduce their vehicles' CO_2 footprint by at least two grams per kilometer. In addition, exact map data and real-time information will make the driving experience more enjoyable and safe by providing useful information about road congestion so that drivers can predetermine alternate routes, as well as a way to adapt LED headlight functions to the road ahead. Last but not least, this solution is paving the road for an even more exciting future where highly automated driving will be a part of everyday life, projected to start at the latest by 2020. An extended horizon enables the drivers or vehicles to adjust driving styles knowing what's beyond the visible horizon leading to improved cost efficiency by reduction of fuel consumption (2–3 %). It lowers emissions by improving engine strategy and economic driving. At the same time, it maximizes driving comfort through adjusted routes based on real-time traffic information. Another interesting application can be found in the optimization of the engine management system in hybrid vehicles as well as range maximization of electric vehicles. Both HERE's 3D route profile and dynamic information about the traffic situation or weather along the way, which may influence the range of electric vehicles, are assisting in this respect. In addition, hybrid vehicles could drive even longer without assistance from the internal combustion engine if the vehicle has dynamic traffic information.

4.3 Smarter Engineering

Telecommunication networks are premier examples of connected devices collaborating together for a common task, whether voice communication, file sharing, or streaming video viewing. As telecommunications users exploit the connectivity for a wide variety of use cases, the networks become increasingly sophisticated in dealing with variations to usage and loads. A stadium filled with tens of thousands of fans can rapidly move in unison from one mode of communication to another in short term periods. Depending on

[6] Pino Bonetti, "next a car that looks around corners, HERE 360, http://360.here.com/2014/01/14/next-car-looks-around-corners/

the event dynamics, thousands of fans could be tweeting, thereby overloading Twitter connectivity one minute and streaming video from a different site in the next minute while watching replays. A complex configuration of network elements enables the telecommunication service provider to respond to fluctuating usage demands and continue to offer a five-bar service to stay ahead of competitors. For most of the service providers, sudden changes in demand are best met with excess capacity, which implies extra capital cost for buying and equipment needed for peak capacity. The alternative is a degradation of performance during peak usage, as many households and hotel residents experience while prime time video streaming using cable modems.

As network elements become sophisticated, they continue to evolve their software. With rapid introduction of cloud technologies, these networks are now moving towards cloud-based networks, where the software defines the network and can be implemented using commodity hardware available in a cloud and can rapidly scale to user requirements by adding more hardware for high demand capabilities. Network Function Virtualization (NFV) and Software Defined Networks (SDN) are two related technologies, which are systematically bringing innovation to drive dynamic configuration, while reducing overall cost to the network service providers, and increasing the network quality for their subscribers.[7] Network service providers use a combination of sensor data, and external data sources to identify demand changes and can use their newly designed cloud-based NFV/SDN supported components to switch configurations accordingly. These technologies offer significant network performance improvement at lower capital cost. However, they rely on Cognitive Things for valuable data to identify changes in network usage.

Most digital products are software-configured. If properly engineered, these products can be easily changed reflecting user preferences, features often used, recent usage patterns, and any other customizations programmed by the engineering teams. Cognitive Things provide plenty of usage feedback to drive engineering changes. In the past, the customization was difficult for hardware design. However, 3-D printing is now making it easier to customize hardware. Let me illustrate an example from automobile engineering where hardware design is rapidly moving to mass customization. Automobile bodies have been a design focus area for many decades. Since the introduction of the Model T by Ford, automobile manufacturers have taken pride in establishing standard car bodies, which take a fair amount of effort to design, but

[7] Prayson Pate, "NFV and SDN: What is the difference?", SDX Central, March 2013, https://www.sdxcentral.com/articles/contributed/nfv-and-sdn-whats-the-difference/2013/03/

can be assembled rapidly using an assembly line and replaced by body repair shops. What would happen if I open up a pizza place and go to buy an automobile to deliver pizza? More than likely, I would buy a four-door passenger car, and use the back seats for stacking pizza. The car will never have a need to carry passengers and yet it has all necessary features for passengers. Can we do better?

Local Motors offers 3-D printing services for automobiles. It offers a design and engineering process for automobiles, which can be used to create purpose-driven custom designs. They used a five-phase process to produce a car for Domino's Pizza with 35 cubic feet of pizza space, which is equipped with an oven capable of keeping the pizza warm up to 140°F. The first phase was to design the Ultimate Delivery Vehicle (UDV) for Domino's Pizza that will revolutionize pizza delivery forever in the United States. The second phase was conducted as a collaborative challenge, where participants worked together towards a common goal. The goal was to select an existing vehicle platform and then how it would be customized for pizza delivery. The third phase was to transform the Chevy Spark into the Ultimate Delivery Vehicle by redesigning the interior to efficiently carry drivers, pizzas, condiments, and other Domino's menu items to customer delivery locations. The fourth phase was to finalize the overall design of the Ultimate Delivery Vehicle as it moves into the surface-modeling phase. The fifth and final phase was to build a prototype of this extraordinary vehicle. [8,9]

One of the primary features of this purpose driven functional vehicle is an onboard oven located inside the modified rear door on the driver side. It keeps pizzas piping hot for those extra long deliveries in sub-zero temperatures. The rear seats have been removed entirely to maximize cargo space for larger orders. These cars are now being produced using Local Motor's 3-D printing facilities.

Can a 3-D printed car be driverless and cognitive? Olli is the first driverless vehicle to utilize the cloud-based cognitive computing capability of IBM Watson IoT to analyze and learn from high volumes of transportation data, produced by more than 30 sensors embedded throughout the vehicle. Using the Local Motors open vehicle development process, sensors will be added and adjusted continuously as passenger needs and local preferences are identified. Furthermore, the platform leverages four Watson developer APIs—Speech to

[8] Domino's Pizza, "Domino's DXP—Ultimate Pizza Delivery Vehicle", YouTube, Feb 22, 2016, https://youtu.be/Rq_YBk8pRL4

[9] Design Process by Local Motors, also available on their website https://localmotors.com/dominos/

Text, Natural Language Classifier, Entity Extraction and Text to Speech—to enable seamless interactions between the vehicle and passengers.

Passengers will be able to interact conversationally with Olli while traveling from point A to point B, discussing topics about how the vehicle works, where they are going, and why Olli is making specific driving decisions. Watson empowers Olli to understand and respond to passengers' questions as they enter the vehicle, including about destinations ("Olli, can you take me downtown?") or specific vehicle functions ("How does this feature work?" or even "Are we there yet?"). Passengers can also ask for recommendations on local destinations such as popular restaurants or historical sites based on analysis of personal preferences. These interactions with Olli are designed to create more pleasant, comfortable, intuitive and interactive experiences for riders as they journey in autonomous vehicles (Fig. 4.1).

You will see more and more mass customized automobiles on the road. Carefully designed probes will collect usage experience from existing automobiles, providing the much needed input to a rapid design process that can produce customized cars, which are far more suited to the specific purpose of a car. Often, these cars may carry alternative designs, which will field-tested and chosen based on feedback from the field.

Fig. 4.1 Ollie: The first driverless vehicle to integrate IBM Watson (Rich Riggins/Feature Photo Service for IBM) (https://www.flickr.com/photos/ibm_media/27593103462)

In a classic engineering process, user feedback is painstakingly collected and collated using large armies of engineering personnel. The product design process for a product is slow and deliberate. While the engineers have the field data, they often must use their gut instinct to decide how their new product idea will fare in the market place. Many new engineering processes are in the works to break this long lead time process, and Cognitive Things will accelerate many new engineering processes, which bring product usage to the center stage in the engineering process.

IBM is introducing "Design Thinking", a revolutionary new process for engineering design, which is transforming IBM's approach to product design. I have participated in this process several times, both for internal product development as well as for IBM clients. It's a framework for teamwork and action. The process focuses on user outcomes, employing an interdisciplinary team to design the product, and enforce relentless reinvention through prototyping. Across corporate America, there is a rising enthusiasm for design thinking not only to develop products but also to guide strategy and shape decisions of all kinds.[10]

In these design tasks, IBM uses design thinking to form intent by developing understanding and empathy with product users. As engineering organizations speed the data collection and product engineering processes, they now have the opportunity to make rapid demand-driven changes to the design, use field experimentations to try out many designs, and optimize configurations based on feedback from the field. Cognitive Things offer us the ability to collect valuable usage data on current products as well as prototypes of the new design, compare and contrast the experience, and use the usage experience to reinvent the product.

4.4 Contextual Marketing

In recent years, there has been a proliferation of big data sources and advanced analytics techniques for behavioral targeting. In particular, usage and location data from the Internet of Things are getting special attention. Network probes have opened up access to detailed usage data from wireless networks. Wireless service providers have access to many high-velocity data sources. They are interested in utilizing this data for themselves and for third-party organizations. Collating all the network data and acting on useful patterns at the right

[10] Steve Lohr, "IBM's Design Center Strategy to Set Free the Squares", New York Times, Nov 15, 2015, http://www.nytimes.com/2015/11/15/business/ibms-design-centered-strategy-to-set-free-the-squares.html

time can present major opportunities for many tier 1 telecommunications wireless service providers.

Etisalat, the leading telecommunications provider in Dubai, analyzes the location data of its subscribers to understand their mobility patterns. In addition to the mobility data, Etisalat also observes their subscribers' media viewing habits and, after obtaining their permission, they are able to track and offer campaigns for various marketers. Working with local retailers, their interest is in understanding the customers at their malls and making context-specific micro-offers based on mall visitors' mobility patterns. The scenario below shows micro-campaigns triggered, based on mobility patterns, subscriber interest, and current location (Fig. 4.2).[11]

Broadcasting marketing campaigns to a wide-ranging market segment can be expensive, and these campaigns often result in poor yield. Many consumer organizations are moving toward smart campaigns for targeting micro-segments. They target a specific set of customers, personalize their campaigns, and obtain feedback from customer responses about how to improve campaign effectiveness. A 2010 study by the Network Advertising Initiative led to three major conclusions. First, advertising rates are significantly higher for behaviorally targeted (BT) advertisements. The average price for BT advertising is just over twice the average price for run of network (RON) advertising.

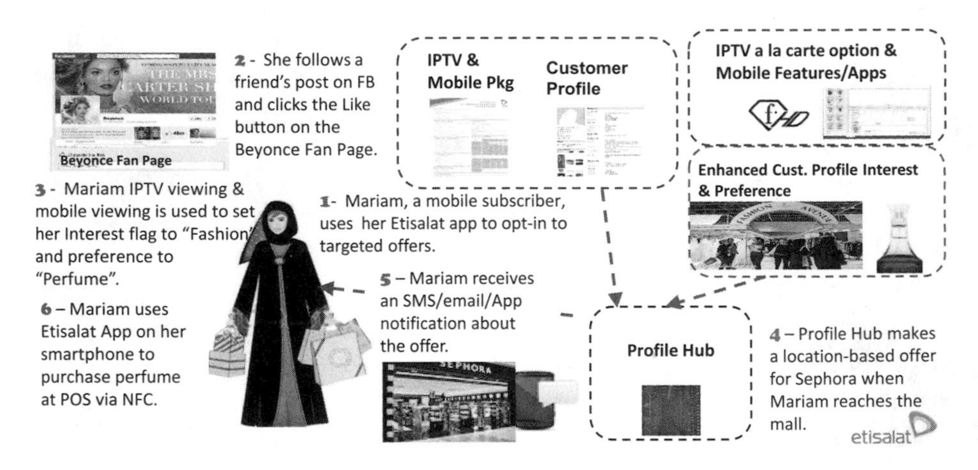

Fig. 4.2 Etisalat location data monetization

[11] Mohamed Hashem, Sambit Sahu, Arvind Sathi, "Creating and Monetizing a Customer Profile Hub—The Etisalat Story", presentation at IBM Insight 2015 Conference, Oct 2015, http://www.slideshare.net/arvindsathi/session-2183-profile-hub-the-etisalat-story

On average across participating networks, the price of BT advertising in 2009 was 2.68 times the price of RON advertising. Second, advertising using BT is more successful than standard RON advertising, creating greater utility for consumers and a clear appeal for advertisers. Conversion rates for BT advertising are more than twice the rate for RON advertising. Third, the majority of network advertising revenue is spent acquiring advertising inventory from Web content and service providers, making BT an important source of revenue for publishers as well as advertisement networks.[12]AU: Sentence: 'Broadcasting marketing campaigns ... expensive for marketing organizations ... ' . I am wondering if you mean marketing departments within organizations – please amend if appropriate.removed

Mobile advertising is a big business, and offers a good channel for BT. In the US, advertisers spent $28.72 billion to reach their targets on mobile devices in 2015, eMarketer estimates, an increase of 50 percent over 2014 spending levels. But—as is common with so many digital advertising channels that offer the promise of measurability and ever-increasing efficacy—performance measurement is still a challenge.[13] To maximize campaign effectiveness, the campaign management system should find the micro-segment and personalize the campaign based on individual preferences.

With the growth of Cognitive Things, behavioral targeting opportunities and concomitant challenges are expected to grow. Automation for such opportunities needs to take into account two specific challenges. First, data flow should allow for the high-volume tsunami of data received at extremely high velocity. The architecture should use advanced techniques for filtering and focusing on data that meets specific conditions. The second challenge is that the intelligence gathered through historical analysis must be shared across the elements to learn from environmental changes and feedback from past actions. Machines need to learn to share intelligence to mimic human learning.

4.5 Proactive Customer Care

Consider a scenario in which a telecommunications customer changes his residence, only to find poor network quality in the new neighborhood (see Fig. 4.3). As he decides he needs to change carriers, he ends up calling his

[12] Howard Beales, "The Value of Behavioral Targeting", Network Advertising Initiative, 2010, http://www.networkadvertising.org/pdfs/Beales_NAI_Study.pdf

[13] eMarketer, "Measuring Mobile Effectiveness Still Challenges Marketers, eMarketer.com, July 2015, http://www.emarketer.com/Article/Measuring-Mobile-Effectiveness-Still-Challenges-Marketers/1012797#sthash.4HFgSapJ.dpuf

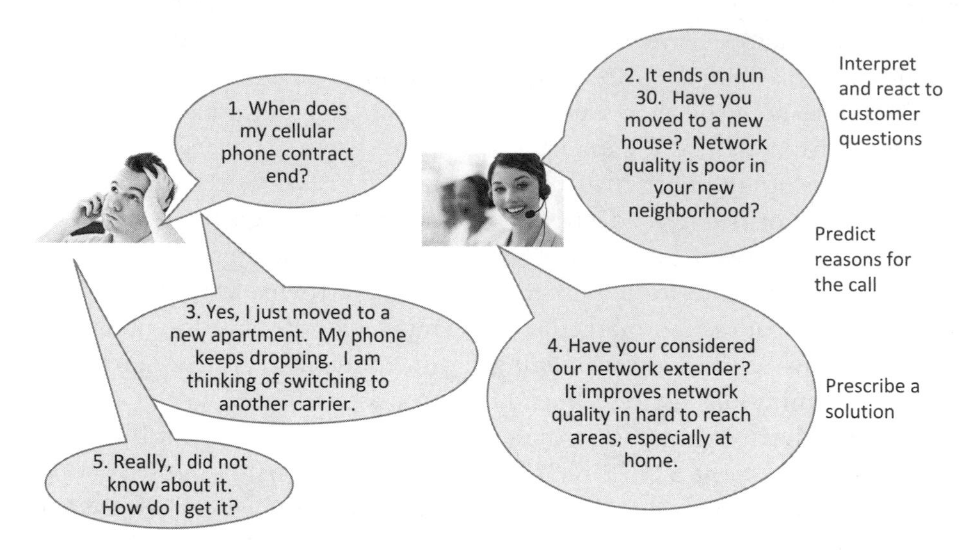

Fig. 4.3 Proactive care scenario

current provider to find out how soon he can discontinue service on his current contract. As the agent goes through his account, she discovers this customer has moved to a new location, possibly his place of residence and has been receiving poor service in the new residence. She checks if he would be interested in a product that will extend the network using an extender, which improves network performance in the hard to reach areas. As he is otherwise satisfied with his service provider, the customer decides to use the network extender, instead of switching to a competitor.

Would it be possible for a digital agent to provide this interaction, combining three skills—ability to engage the customer in a dialog, ability to predict reasons for the call, and ability to access and execute a prescribed solution?

If the wireless service provider is able to identify a new hangout for a valuable customer and knows the customer is now getting a poor service, should they wait for the customer to call? Most likely, the customer may never call and churn. If the customer calls, what is the ideal call flow? Today, most service providers wait for an irate customer to call and often respond to the question without digging deeper into the reason for the call. Can we take the customer care experience to a proactive level? The smartphone provides a direct link between the wireless service provider and the customer, and using its app notifications, the digital agent can engage with the customer even before they call the service provider.

While I have used natural language to describe the scenario as a call center call, it can easily be recast as a session on the service provider's website or smartphone app, where the exchange could just as well be menu and graphics driven. The key aspect of the scenario is the agent's ability to use sensor data to identify the real reason for the customer distress and to engage in a meaningful conversation about network quality. It may also be possible that the first such conversation happens between a human agent and the customer, and the digital agent is just an observer noting down all the successful interactions. As the next customer calls with the same pattern of care issues, the digital agent learns from and uses techniques that the human agent has successfully employed to turn the problem into an opportunity. The cognitive agent learns through the success and failure of past interactions, and continues to refine the prescribed processes based on field success.

Customer care is often achieved via call centers, which is the interaction of last resort. Often using outsourced agents in offshore locations, organizations manage large volumes of calls from their customers. Due to high agent turn over, it is hard to find and retain good agents, and care organizations find it hard to build and retain good prescribed processes. Using call center logs, interactions data from other channels and customer usage information, a cognitive agent can piece together proactive care patterns and gradually move the experience to a more positive, as well as efficient, channel. In many cases, new digital channels are providing both efficiency as well as quality of interaction, and can significantly improve customer experience while reducing channel costs.

A hefty percentage of subscriber churn can be experienced when subscribers struggle with their malfunctioning devices. Instead of seeking a replacement device, they replace their supplier. In order to reduce churn, service organizations often replace a device perceived to be malfunctioning, where the device problems cannot be tracked or fixed. What if the problem with the service had nothing to do with the device? The customer walks home with a new device, which shows the same "malfunction", and the service provider faces returns of products that are not really faulty. Is this a problem with a product or with the diagnostics process?

As products become increasingly sophisticated, they are in need of a complex and cognitive diagnostics process—one that uses a fair amount of field sensor data, learns from and mimics the best service professionals, and evolves with changes to the product and its environment. Most decision-tree-driven diagnostics processes do not adequately cover the complexities of a product diagnostics, and often lead to dissatisfied customers and "no fault found" product returns. A major benefit from proactive care is the ability to pinpoint the real reason for a performance degradation and to proactively take care of it consistently, before it becomes a source of irritation to a group of customers.

4.6 Counter Fraud Management

Counter fraud management teams work towards isolating and shielding service providers from fraudsters who steal services and conduct illegal transactions. A very small percentage of transactions are fraudulent. However, according to the Communications Fraud Control Association (CFCA), they add up to $5.22 billion in annual losses for communications service providers alone.[14] Unfortunately, many industries—credit cards, automobile insurance, health insurance, to name a few, are seeing increasing fraudulent use through identity theft. While Counter Fraud Managers find ways to protect the organization from known fraudulent patterns, fraudsters keep finding new loopholes in processes and systems to invent new ways of defrauding people. How can a fraud management function make agile changes to its fraud detection and prevention software to effectively find and deactivate fraudulent services before they cause excessive revenue leaks?

Cognitive Things provide useful ways to catch the thieves. By trapping all usage, Cognitive Things can provide a counter-fraud department with meaningful observations, where the product usage is inconsistent, or in some cases, illogical. Consider the example of a fraudster stealing the identity of a smartphone to make fraudulent purchases thousands of miles away from the physical location of the phone. A counter fraud system can pinpoint where the customer seems to be present in two locations, simultaneously. Using location and usage information, transactions can be monitored against a digital signature—a customer identity based on typical usage and mobility patterns. As more Cognitive Things are integrated, these signatures become increasingly sophisticated, making it difficult for a fraudster to replicate the signature.

4.7 Chapter Summary

In this chapter, we have looked at a number of ways in which Cognitive Things are contributing towards organizational functions, including operations, engineering, marketing, care, and counter fraud management. In each case, I used a series of examples to show how Cognitive Things support the function by collecting data, collating it as needed for the specific function, developing insights, and offering ways in which this insight can be applied to improve the

[14] 2013 Global Fraud Loss Survey," Communications Fraud Control Association. Note: registration is required to obtain a copy of the survey report. http://www.cfca.org/fraudlosssurvey/

function. In the case of smarter operations, Cognitive Things collect useful operational data about the organization, and analyze this data at key decision points to facilitate operational decisions such as placing grocery clerks at point-of-sale or providing better road condition management for a city. The product usage data can be used for engineering a better product, whether it be an improved network configuration for a telecommunications network, or automobile design for a car. Data about location and product usage can be used in contextual marketing to customers, thereby providing meaningful promotions at the right time. Organizations can employ Cognitive Things to predict failure and to direct proactive care for their customers. Last but not the least, Cognitive Things can be used for detecting fraudulent behaviors and for preventing revenue loss due to fraud.

In the next chapter, we will see how this data can also be reused externally and can be monetized and sold to others.

5

Reuse and Monetization

5.1 Introduction

In 1989, the movie *Back to the Future II* depicted a time traveler moving forward in time to 2015 and marveling at the new inventions. When 2015 arrived, movie fans compared the present with what they had seen in the movie in 1989. While the movie had erred with regards to transportation in its depiction of flying automobiles, its projection for 2015 included weather prediction and accuracy improvements. The story writers got this one right, though they probably never guessed how weather prediction would take such a massive leap. If they had guessed correctly, the movie would have shown many Cognitive Things supporting the data collection for weather prediction.

As Cognitive Things collect and collate data, they become important data sources for cloud data services. Weather is a prime example of cloud data services, which reuse data and monetize it for a large number of new consumers. As monetization progresses, it brings new sources of revenue, thereby subsidizing the infrastructure for data collection and collation. The end result is a subsidy for Cognitive Things by consumers of derived data.

Information about the weather is an obvious example of something worth sharing. In any culture, weather is the most common small talk subject. No other type of information collected is as clearcut. A number of IoTs collect location information. However, the location of an individual is far more private as compared to the weather. At the same time, aggregate location data for a particular location could be relatively easier to share and still provides meaningful data about movement in a community. Media viewing is another commonly measured and shared parameter. In this case, advertisers have used

A. Sathi, *Cognitive (Internet of) Things*,
DOI 10.1057/978-1-137-59466-2_5

media viewing as a monetization to subsidize media publications. Viewers are used to viewing advertisements in return for subsidized media. As the World Wide Web gained popularity, it used this subsidy as a basis for a number of media related functions. Most email subscription plans offer a free email service, in exchange for advertising. As Cognitive Things are becoming popular, so is the interest in using the usage data as a means for subsidizing the use of Cognitive Things. For example, many fitness apps provide a free service to their subscribers with a business model to use the subscriber and usage data for marketing related products and services.

In this chapter, we will look at a number of case studies covering different aspects of reuse and monetization. First, we explore some of the common data sources that are being monetized—weather services, media viewing, and location—as well as the players involved in data consumption. Next, we look at the agents of monetization—the data producers, the consumers, the market place, and the organizers who govern and organize the data, in exchange for part of the monetization revenues. Finally, the challenges that make monetization difficult will be discussed, together with the cognitive functions needed to improve the monetization business models.

5.2 Weather Services

Weather storms are tricky. Good prediction must accompany good planning to deal with exceptional situations. Atlanta is a city that hardly ever witnesses snowstorms, so the city comes to a standstill on the rare occasion when a snowstorm actually happens. It had good weather prediction information just before a terrible storm hit the city in January 2014. The city decided to close all schools and businesses in preparation for the weather. Unfortunately, this mass simultaneous closing resulted in the worst ever traffic jam in the city as the snow froze converting the streets into ice skating rinks and nearly every one at that time was on the road at the same time trying to go home. The mayor admitted the mistake and learned an expensive lesson for the future, which is how to plan for a storm and release the traffic in stages to allow everyone to get home without a major traffic jam in the worst of the weather.[1]

When the rain falls and the wind howls, shoppers may cancel or postpone shopping trips to avoid braving the elements. Accordingly, atmospheric conditions can mean the difference between a robust shopping season and one that

[1] Jason Hanna and Ralph Ellis, CNN, "Atlanta mayor: Early city exodus due to snow crippled traffic", CNN, January 29, 2014, http://www.cnn.com/2014/01/29/us/atlanta-mayor-reed-weather/index.html

falls short of expectations. Moreover, weather can disrupt suppliers, shippers, and other components of the retail value chain, making it hard to serve customers who still came out despite the weather. Weather can even affect business by triggering subtle shifts in customer sentiment, as when prolonged heat waves increase general feelings of listlessness and fatigue. The Weather Company has provided many interesting examples of the influence of weather on consumer behavior. The conventional wisdom is that ice cream sales should increase as the weather gets hot. However, the sales of supermarket ice cream actually drop when the temperature exceeds 25 °C (77 °F), because shoppers fear that it will melt on the way home. Barbecue sales tripled in Scotland when temperatures rose above 20 °C (68 °F). In London, however, the figure is 24 °C (75 °F).[2]

I was fortunate to attend a presentation by David Kenny where he covered the fascinating story of the Weather Company last year. IBM was so impressed with their platform, that they ended up acquiring the assets in October 2015, and now David is the General Manager of Watson products at IBM. He explained to us how airlines are both consumers and users of weather data. When airplanes fly at 30,000 feet, they must control the aircraft when flying in air pockets, as well as manage the safety and comfort of their passengers. As an airplane travels through an air pocket, the information collected is valuable to all the other airplanes that are traveling on the same path. Each airplane collects and processes air pocket information and submits its observations to the Weather Company. The Weather Company processes the data collected by all the participating airplanes around the globe, organizes it and makes it available for consumption. The airplanes are the direct beneficiaries as they can receive the information in advance of reaching an air pocket. In conducting this information integration, both airplanes as well as the Weather Company are sensing, filtering, aggregating, and reshaping the data to make it usable by others.

Additionally, the information collected from the airplane allows the Weather Company to map the 63 vertical miles of atmosphere more precisely. Without the airplane data, it would have been much harder to map wind direction. The Weather Company has also expanded its reach to many other data points. In the past, weather information was primarily based on 8000 data points created by the various governments around the globe. Over time, more than 172,000 personal reporting stations started to contribute to the weather data. As The Weather Company gets better at collating data from

[2] James Kobielus, "Weather data analytics: Helping retailers predict and meet customer demand", September25,2015,http://www.ibmbigdatahub.com/blog/weather-data-analytics-helping-retailers-predict-and-meet-customer-demand

many sources—airplanes, connected cars, buildings, and so on—it can slice and dice this data across 2.2 billion locations. Today, it serves an average of 15 billion forecasts a day with a peak of 26 billion forecasts.

The weather information can now lead to many new innovative services. Consider the impact of hail damage on vehicles. I used to live in Denver where hailstorms were a frequent occurrence, showing up with large hailstones capable of creating big dents in the roof of my car. I talked to my car dealer who complained that a hailstorm destroyed his entire inventory one afternoon. If only there had been prior warning, he could have moved all the cars to a safe location and avoided incurring such colossal damage. While consumers and businesses check weather, a proactive notification of weather has enormous value to insurance companies, which have to foot the bill for hail damage. They can both avoid the hail damage by proactively notifying their customers and can also increase their call center capacity to handle additional calls for processing inclement weather driven insurance claims.

Weather forecasters have a difficult job. Their predictions are validated by the actual occurrence, thereby providing a strong feedback loop on the quality of weather prediction. Crowdsourcing from a large number of sources provides a lot more source data. Organization and refinement of this data is equally important. As shown in the movie *Back to Future II*, the information is accurate and in real-time, providing instantaneous access to its crowdsourced data, sorted and organized by a world-class meteorological organization.

5.3 Media Viewership Services

As commercial newspapers and magazines evolved, they began to use advertising as a source of revenue, thereby subsidizing the subscription cost for their audience. The television industry followed the same path and primetime slots on national television started to carry big price tags. For a number of decades, television producers relied on a control sample of audience viewing habits to gauge the popularity of their television shows. This data was collected at first using extensive surveys in the early days of television programming and then, later, special devices placed on a sample of television sets. Nielsen initiated its television monitoring service long before Internet of Things was on the horizon. Its monitoring unit replaced error-prone manual reporting and provided a mechanism to collect audience behavior from a small, well-defined sample of viewers.

With the advancement in the cable set-top box (STB) and the digital network supporting the cable and satellite industries, cable operators can now collect channel surfing data from all the STBs capable of providing this information. As a result, the size of data collected has grown considerably, providing marketers with finer insights not previously available. This information is valuable because it can be used to correlate channel surfing with a number of micro-segmentation variables.

Television STB data is available at the household level, while mobile device content data allows for content viewing by individuals. Consumers are beginning to watch content on both platforms, and sometimes they even use both at the same time for complementary information. This data is collected from the devices, cleaned up, and correlated with programming data to ascertain the timing of customer behavior. If I use a two-way STB to watch television, the supplier has instant access to my channel-surfing behavior. Did I change the channel when the advertisement started? Did I turn the volume up or down when the commercial started to play? What if the consumer started to watch the television, but left it on for the day while going to work? A fair amount of cleanup is needed before this data can be analyzed. STBs are geographically located. If we know the television viewing habits of a community of people, that information can be utilized for beaming specific messages to that community. Aggregation and correlation can be used to analyze STB location data, combined with STB usage data.

Today, with the availability of cable viewership data from cable set-top-boxes (STBs), cable operators are aspiring to supply their television viewership data to marketers. In the USA, their first major attempt was Project Canoe, which was formed as a consortium of six cable operators. This was in response to a Google–Echostar partnership, which used satellite viewing on the Dish network.[3] Unfortunately, the Canoe project failed in meeting its original goals. Six years later, the effort was scaled back to providing interactive advertising for video-on-demand. While Canoe was not able to get all of its constituents on the same page, it did in some ways show proof of concept. It released the results of a study jointly conducted with the American Association of Advertisers, in which a panel of 4,200 cable subscribers revealed increased product acceptance when shown interactive ads from brands like Honda, Fidelity, GlaxoSmithKline, and State Farm. According to the year-long

[3] Erick Schonfeld, "Project Canoe: Cable companies paddle to catch up to Google in targeted TV ads," TechCrunch, March 10, 2008, http://techcrunch.com/2008/03/10/project-canoe-cable-companies-paddle-to-catch-up-to-google-in-targeted-tv-ads/

study, 19 percent of adults 18–49 said "yes" to interactive offers, while 36 percent expressed a likelihood to purchase.[4]

The interactive video market has grown in the meantime, leading to yet another avenue for media and advertising research. Unlike linear television (which is shown on airwaves), interactive television, often termed "nonlinear," is well instrumented. Video content managers keep track of viewing details for each of their subscribers. As the nonlinear content is becoming increasingly mobile, the location of the subscriber may no longer be static. Last but not least, interactive viewing is done more often by an individual rather than a household. In my own house, interactive video has significantly reduced our traditional television viewership. The Canoe project finally found its sweet spot and has survived and thrived as a supplier of interactive television research data.

The online viewership of data and video is closely scrutinized. Data management platforms (DMPs), such as Blue Kai (now a part of Oracle), Adobe, Aggregate Knowledge (now a part of Neustar), CoreAudience, Knotice (now a part of IgnitionOne), and nPario, track and analyze a fair amount of data from Internet viewers. Marketers are very interested in understanding a cross section of viewers and a holistic viewership data, including social, mobile, display, and search. Potentially this data can then be combined with the delivery of messages via targeted advertising or in other ways. With the rise in the mobile platform for viewership, this data can also be correlated with location data to understand and mine geographic differences.

5.4 Location Services

Mobile platforms have invaded the shopping process and now provide great options for additional information access at shops. These location-aware phones can be pinpointed within 20 yards of accuracy in a shop offering Wi-Fi service, and marketers have provided additional information about products, which can be accessed on the phone. Shoppers also use their phones for comparative shopping, seeking advice, comparing prices, and looking for deals from the competition. In response, stores are actively using phones for location-aware marketing, mobile coupons, loyalty cards, and related marketing activities. Since a mobile device is location aware, it can be used for active placement of location-aware marketing messaging. Also, compa-

[4] Daniel Frankel, "Why Canoe abandoned ship with interactive TV ads," Gigacom, Feb 22, 2012, https://gigaom.com/2012/02/22/419-why-canoe-abandoned-ship-with-interactive-tv-ads/

nies are using mobile devices to develop interactive marketing programs. For example, Foursquare (www.foursquare.com) encourages me to document my visits to a set of businesses it advertises. It provides me with points for each visit and rewards me with the title of "Mayor" if I am the most frequent visitor to a specific business location. Every time I visit Tokyo Joe's, my favorite neighborhood sushi restaurant, I let Foursquare know about my visit and collect award points. Presumably, Foursquare, Tokyo Joe's, and all the competing sushi restaurants can use this information to attract my attention at the next dining opportunity.

We carry our cell phones everywhere, and have now started to use mobile devices to watch movies, browse social media, and make transactions. How can a marketer collect, organize, and analyze location data? A cell phone is served by a collection of cell phone towers, and its specific location can be inferred by triangulating its distance from the nearest cell towers. In addition, most smartphones can provide Global Positioning System (GPS) location information that is more accurate (up to about 20 meters) but can rapidly drain the cell phone battery. In most marketing situations, cell tower location data combined with occasional GPS is good enough. The location data includes longitude and latitude and, if properly stored, can take about 26 bytes of information. If we store 24 hours of location data for 50 million subscribers at the frequency of once a minute, the data stored is about 2 terabytes of information per day. This is the amount of information stored in the location servers at a typical tier 1 wireless service provider. While that is a lot of data, it can be rapidly aggregated to keep only meaningful information.

By itself, longitude-latitude data is hard to analyze. I may need to know the granularity of the data and a measure of proximity so that I can infer whether a person is at one location or another. If I have to count the number of people sitting in a building, we need simple measures for location, which can be counted. While subscribers move around a lot, they are still creatures of habit. A small number of hangouts dominate their locations—possibly home, work, and social meeting places. Location and usage data can be used to establish the identity of an individual based on specific hangouts visited by that individual and specific usage of their phones in those locations. This insight can be used to identify someone who switches brands regularly and can be used to provide incentives for them to stay with a specific brand.

Location data can be generated at different levels of accuracy. Typical cell tower data as described above is accurate within 1–2 kilometers. However, wireless subscribers often turn on GPS to find directions on their cell phones. At the expense of a cell phone battery that may be rapidly consumed, the location data captured through GPS can be uniquely represented using an 8-byte code and provides accuracy of 20 meters. Similar accuracies can be achieved when consumers use Wi-Fi in a sports stadium. Wi-Fi location data has one more advantage—in addition to the fact that it does not overdrive battery consumption, it also improves our ability to connect to the Internet. A public gathering area like a stadium may offer free Wi-Fi to its audience to ascertain their location data and use it for a variety of operational and marketing purposes. For example, the stadium may offer a visitor advice on which gate to use for entry to the stadium based on current visitor location, seat location, multiple gate locations, and the length of the lines at each gate. There are many interesting marketing opportunities once we have a person with a smartphone located in a stadium who is able to watch the television screen, interact with a little screen, and has a fair amount of interest in buying merchandise located around him/her. By combining more aggregated cell tower data with Wi-Fi data, we can now combine the behavioral characteristics of a shopper (couch potato vs. frequent mall shopper) with the shopping behavior of the shopper (time spent in each aisle or combinations of aisles visited). A savvy data scientist can also use clustering algorithms to establish micro-segments by finding individuals who follow similar mobility patterns. Some of the micro-segments are based on people traveling to similar locations. However, more complex micro-segments are based on mobility patterns to diverse locations. For example, a statistical program can find active weekend golfers who wake up early on the weekend and show up at the golf course for a Saturday morning game. These golfers may be showing up at different golf courses around the globe, but share the Saturday morning mobility pattern. This micro-segment is of enormous interest to golf companies, golf resorts, and the leisure travel industry.

5.5 Location Data and Contextual Marketing

The prepaid wireless market is fairly competitive in the growth markets, where wireless providers sell SIM cards and devices belonging to a subscriber can use more than one SIM card from competing service providers. In most situations, consumers buy their mobile device directly from cell phone manufacturers, while telecom providers sell subscriber identity module (SIM) cards to enable these devices to use their network. If a consumer runs out of SIM card minutes, he looks around for a SIM card retail outlet and buys a new

SIM card, which might possibly belong to another telecom provider, thereby resulting in brand switching. However, the first telecom provider has a running balance of remaining minutes for each subscriber and could possibly remind the customer to buy the next SIM card when the balance is low and there is a SIM retail center nearby. An intelligent campaign engine could keep track of location data and remaining minutes, and configure a campaign appropriately at the right moment. As telecom organizations become savvy at running their location aware campaigns, they are collaborating with other marketers to run their campaigns.

I wish my car would do the same. Most of the time, I see a warning for an empty gas tank on my dashboard right after I have passed a gas station. The car has a navigation system that has full awareness of the location of gas stations. It also knows my current location and is able to signal when I am running out of gas. If only the car had an additional capability of combining three data items from two sources, and I would never have to experience tense moments when I am praying the car will not run out of gas before I find the nearest gas station.

There are many consumers of location data including the marketers interested in contextual marketing. How does this contextual marketing work and how is it different from traditional marketing? Customers may have the best intentions to use promotions targeted to them, but may not always act upon those promotions. For example, let us look at coupons from consumer products. Traditionally, these coupons were delivered in a printed form, such as Sunday newspapers, coupon booklets, and so on. Many coupons were printed, but unfortunately were not seen by customers. The precious few coupons that attracted customer attention still had only a small chance of being redeemed. Customers had to cut out the coupon, bring it to the store, find a product, and redeem the coupon. There was a chance of leakage in each of these steps. Electronic coupons started to bridge the gap. Groupon offers coupons that can be organized in a mobile wallet on smartphones, and can be redeemed, thereby reducing the number of items a consumer must carry to the point of consumption. The use of smartphones for grocery shopping has provided grocers with the next level of automation in coupon redemption. If the store offers Wi-Fi, customers walking through the grocery aisles can be tracked, and the smartphone can be used to identify aisles where marketed products are placed. At the end of the shopping trip, the shopper can link the phone to the point of sale, offering the coupons to be automatically uploaded for redemption. Campaign success can be traced across customers and used for fine-tuning the targeting of the campaign. It is possible to trace customers who are not likely to respond to a campaign and improve campaign yield. Also, the impact that a campaign has on customers can be studied using experiment design, as a marketer may test competing campaigns to different subsets of the target market, comparing their effectiveness and choosing the one with the best results.

Intermediaries such as Groupon in the example above, are utilizing any available data to target the customers. They seek location data from a variety of sources, curate this data to identify micro-segments of interest, and offer marketing services that deliver contexual campaigns to targeted customers, matching demand and supply as close as possible. If a marketing service is able to target customers with near empty gas tanks and direct them to a gas station, which is competing with other gas stations in the area, the marketing services company would deserve a monetization fee. In most cases, their service value is measured by the campaign yield. As the marketing service company is able to place a large percentage of coupons that are utilized, it reflects a high campaign yield.

5.6 The New Advertising Market Place

Going back to the media-viewing example, the world of advertising is becoming increasingly mobile, with a very sophisticated network of services to monetize location and viewership data and target customers using this data. In the broadcast era, advertising was concentrated into a couple of media outlets each using a series of direct negotiations. Advertising agencies managed bulk purchasing and used their purchasing power to negotiate the best terms for their customers—the marketers. As part of their services, the advertising agencies supported marketers with media planning, and research capabilities, and thereby provided a one-stop shop (see Fig. 5.1 Since the audiences were concentrated and the messaging was relatively unified, the ecosystem was relatively simple.

Today's viewership and associated advertising opportunities are far more complex (see Fig. 5.2). There are many more media formats. The display and apps vary with the devices. Advertisers have both linear opportunities, which are synchronized with the broadcast and nonlinear opportunities, where the viewership is for a previously recorded broadcast. Direct negotiations and bulk purchasing for advertising spots are being augmented with auction markets.

The online advertising food chain is also becoming increasingly sophisticated. The digital advertising market is rapidly moving toward real-time-bidding or programmatic advertising involving publishers and advertisers that use a complex network of demand-side platforms (DSPs), supply-side platforms (SSPs), and big data-driven data management platforms (DMPs), as shown in Fig. 5.3. Online advertising provides a tremendous opportunity for advertising to a micro-segment and also for context-based advertising. How do we deliver these products, and how do they differ from traditional advertising?

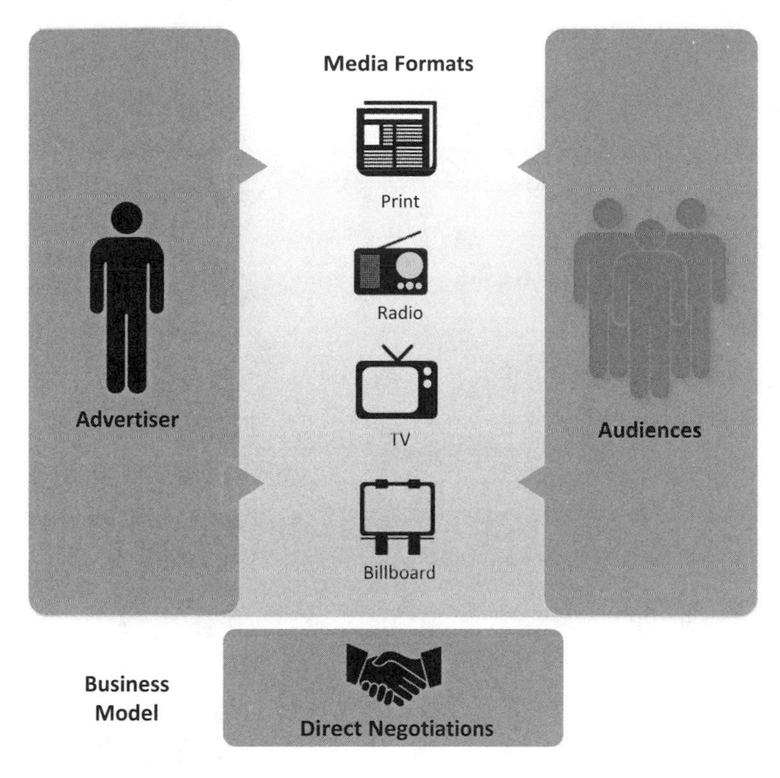

Fig. 5.1 Advertising in the broadcasting era

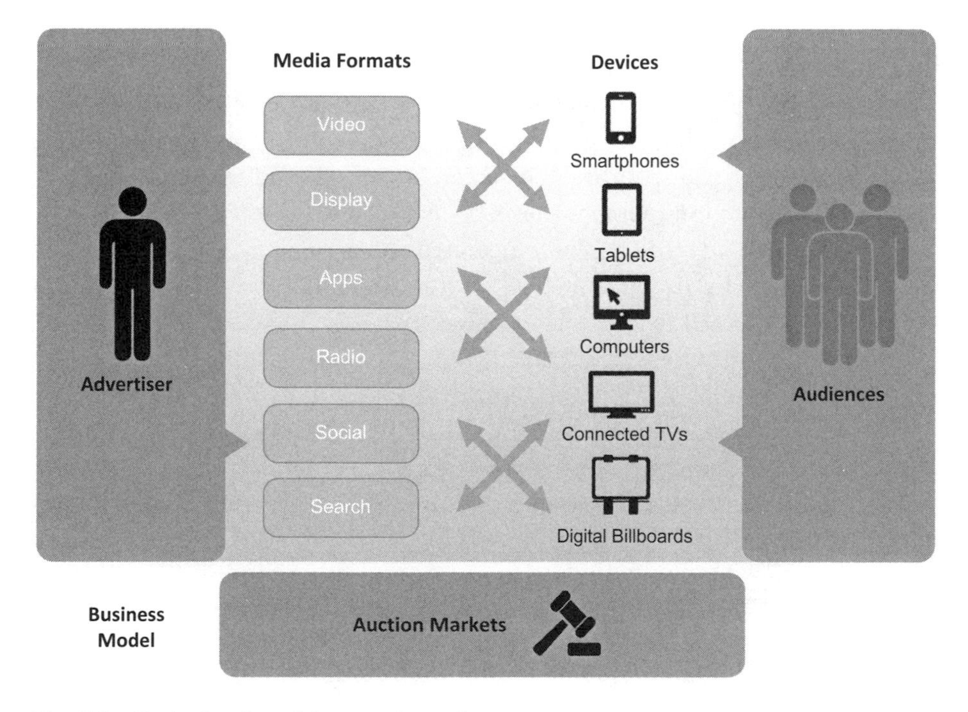

Fig. 5.2 Today's advertising market place

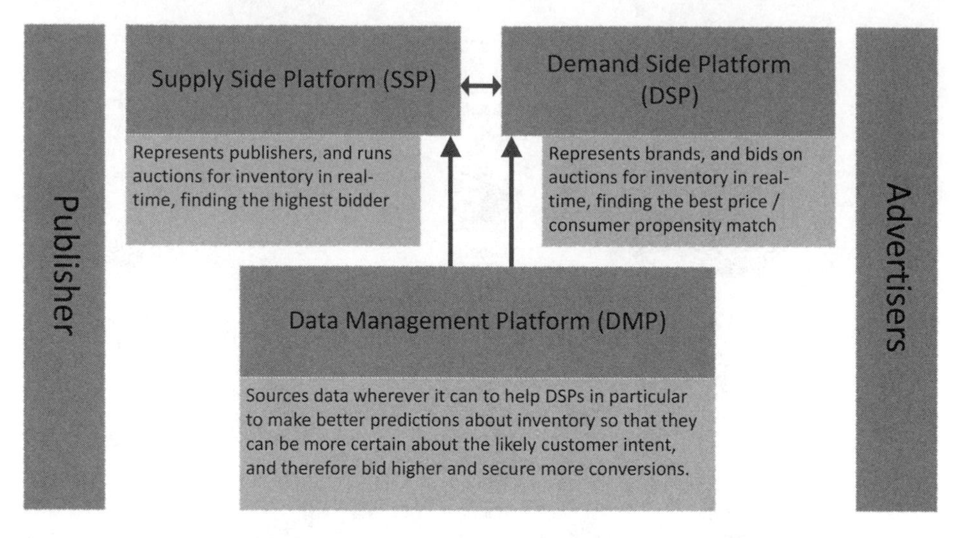

Fig. 5.3 Advertising market participants

Google delivered a major disruption in the advertising marketplace by offering measurements and payments based on advertising clicks. Once an advertisement is placed on a browser screen, the click-rate measures its effectiveness in being noticed by the customer. The next wave of changes came with real-time bidding for advertising. In the online advertising world, publishers such as Google Adwords offer a bidding process in which advertisers bid for placing their advertisement. In less than 100 milliseconds, Google collects a number of bids for each advertising opportunity and decides which advertisement(s) to display on the screen. Once the advertisement is displayed, the user action (namely, the click) is captured and reported.

Programmatic advertising combines a number of these innovations and uses sophisticated Artificial Intelligence tools to place advertising without human involvement.[5] In 2016, US programmatic digital display ad spending will reach $22.10 billion. That's a jump of 39.7 % over last year, and represents 67.0 % of total digital display ad spending in the USA. "Programmatic is extremely efficient and unparalleled in its ability to pair rich audience data with ad inventory and targeting," says eMarketer senior analyst Lauren Fisher. "Buyers and sellers are also becoming more comfortable with the technology. As a result, it is being rapidly adopted across a variety of channels and

[5] Or Shani, "Get With the Programmatic: A Primer on Programmatic Advertising", Marketing Land, August 22, 2014, http://marketingland.com/get-programmatic-primer-programmatic-advertising-94502

ad formats."[6] Unlike the past, savvy marketers can now purchase advertising spots based on what works in the field. The marketers match their advertising to their targeted micro-segments, and use intent-driven information sharing, and fine-tune their messaging using field experiments.

Monetized data from location, usage, media viewing, purchases, social media, and many other sources, can be used to develop a 360-degree view of the consumer. This is where data collected from the Internet of Things can power a cognitive process to uniquely establish a conversation with a specific customer, using the power of digital connection and digital advertising. A marketer can keep track of specific customer needs, information exposure, and attitudes and consult with them to provide additional information without blasting them with repetitive advertising on items that hold no interest for the customer. Digital agents of tomorrow will use this power and the monetized data from a variety of sources to deliver unique and targeted messages, thereby minimizing the cacophony of marketing messaging and improving the yield of advertising.

5.7 Monetization Candidates and Criteria

If weather, location and media viewing are the first three data sources to be monetized, what will be the next three? A number of industries are bracing to use monetization as a driver for their business plans for IoT. Are they all going to be successful, or are some more likely to succeed than others? Drawing on the early successes, what can be established as the criteria for selection?

Almost any data collected by IoT is a candidate for monetization. As cars become better instrumented, the driving data is of interest to manufacturers, insurance companies, and the city planners. As fitness devices collect physical activity, the data is useful to health insurance companies, doctors and employers. As home appliances are connected to the Internet, companies supplying household goods are interested in collecting and dissecting usage data to provide better product offers. The industrial usage is equally interesting. As telecommunications service providers use networks, equipment vendors are interested in understanding usage data to improve their product design and marketing. The factories can provide data to raw material and machinery equipment providers. Banks can use the transaction data to improve their banking services and package this data for their ecosystem. In each case, an established marketplace can use the monetization platform to create new business models for

[6] Media Buying, "More Than Two-Thirds of US Digital Display Ad Spending Is Programmatic", eMarketer, April 5, 2016, http://www.emarketer.com/Article/More-Than-Two-Thirds-of-US-Digital-Display-Ad-Spending-Programmatic/1013789

cooperation across businesses, thereby providing better products and services to their end customers. If improperly run, the same platform can negatively impact core businesses and could easily become a major embarrassment to the organization selling the data as well as to the organization utilizing the data. Weather was a great early success story for a couple of reasons:

1. Everyone likes to get better weather data, rapidly generating a massive user base to consume the data.
2. There are no privacy or competitive issues. If one airline collects this data from their fleet of airplanes, they would share with other airlines.
3. The Weather Company provided a platform for data collection and refinement.

Media viewing and location are rapidly gaining interest, but face major obstacles in their monetization. While consumers are resigned to viewing advertisements in exchange for subsidized media cost, they are resistant to privacy invasion. Also, each media-viewing platform has a biased sample of the population, based on their share of the market, unless they have a monopoly in a geographic area. Therefore, most of the value goes to the platform creator, such as Nielsen or Google. Location data is even choppier. Most of the location data collected by telecom organizations is less granular and most telecom organizations do not have an adequate business model for collecting GPS data, which is also sparse. Telecommunication organizations also faced backlash as they set up business models for data monetization, both from regulators and customers.

In the aptly titled book, *The New Oil: Using Innovative Business Models to turn Data Into Profit*, Arent Van 't Spijker favorably compares the processing of data to oil refinement. In its raw form, data is not very usable, and its value enhances enormously with proper refinement.[7] In most situations, data preparation, curation and analysis are driven by the organization owning the data. As the focus moves from data owners to data users, who purchase the data or related insight, the refinement process must gear towards the needs of the external data users. For example, Nielsen takes a lot of care in data refinement for viewership, as it is the data source and provides a revenue-generating data service to its marketing customers. When the cable operators collect viewership data, it is refined to support their core business, which is to provide quality service to their customers. The refinement process must change radically to support media viewership measurements.

[7] Arent Van 't Spijker, "The New Oil: Using Innovative Business Models to turn Data into Profit", Technics Publications, 2014, ISBN: 978-1-935504-82-5, https://www.amazon.com/dp/B00LMNZ1C6/

5.8 Cognitive Monetization

While the first wave of monetization provided a major thrust to advertising and marketing and provided many free services, including search and email, it has also gathered a fair amount of privacy concerns. While viewers in the past decades complained about the plethora of advertisements on prime time network television, they were relatively content with the privacy management. The new non-linear advertising is not only disruptive to their content-viewing, it is also very intrusive and uses a fair amount of contextual data for targeting advertising. The backlash will definitely rise. On one hand, consumers like targeted advertising covering topics they would like to research, but they are not at all comfortable with full-fledged invasion of privacy when it comes to their email reading, and location data. As I sampled opinions around the globe, there was a full spectrum of responses—all the way from the delight of a teenaged girl who wanted to tell all her friends about Foursquare capabilities for helping them shop, to environments where both public opinion as well as regulations protected individuals' privacy.

Clearly, while monetization sounds simple, it has many layers of complexity, making it hard to work. So, it is a complex business requiring careful consideration and complex rules. Does that sound like a cognitive opportunity? I must admit, this is relatively new ground. There are a couple of observations to ponder:

- Observation 1—As stated earlier, the beauty of data lies in the eyes of its beholder, which is the user in this case. They have specific needs for data collection, refinement, and analysis. Very often, what may be needed is not the most private data, but relevant patterns, which are only known to the user. For example, a quick service restaurant would like to know how many commuters pass by their store to buy from a second store. They are not interested in individual commuter paths, but an aggregation of commuters on a path.
- Observation 2—As devices connect to the Internet, they are capable of overwhelming data networks and cloud data storage worldwide. Operational efficiency will drive intelligence in the edge and combined with observation 1, which means that the rules are created by monetization end-users, but applied as close to the source as possible. Thus, intelligence will flow from users to data sources, and related patterns will subsequently be collected by data sources and sent to the users.

- Observation 3—Data privacy must be organized, both at the group and individual levels. I like to share certain events publicly, other events in closed communities, and some just with my family members. One of my close friends would not share any information, so I can share his picture but cannot tag him. There are other friends who will share minute details about their day-to-day behavior as well as their sentiments openly with the world. I would rather get relevant advertisements even at the cost of my privacy, but many of my friends would never buy a product if the advertisement invades their privacy.
- Observation 4—Data governance must help data users identify data sources, biases, data quality, and other nuances. It is not a good idea to base a decision based on a biased sample covering half the population if the intent of the decision is to have statistical significance for the entire population. For example, a phone sample covering landlines may exclude most of population under age of 30 in US, since most young people do not have landlines. If I were to predict their behavior using a biased sample, it is likely to be worse than a truly random small sample, which may easily provide a quality decision with low sampling error of 2–3 %.

Developing the full potential of monetization requires good collaboration between man and machine to account for these observations. As monetization gathers steam, cognitive functions provided by machines will play major role in managing analytics in the edge, fine tuning privacy and in interpreting data biases.

5.9 Chapter Summary

Monetization is often seen as the major business driver for Cognitive Things. Early success in weather, media viewing and location data monetization has provided others with a glimmer of revenue potential through new business models. This chapter covered a couple of success stories and showed how business models are emerging for data creation, consumption and refinement.

However, monetization is not easy and is not “business as usual”. It opens up the data providers and consumers to customer and regulatory backlash and may significantly increase the cost of data collection and refinement. I have made a number of observations, showing the need for intelligence at the edge and customized privacy policies to account for individual and group preferences.

This chapter concludes the use cases view of Cognitive Things. In Chapter 2, we looked at the definition of Cognitive Things. I described how Cognitive Things are employed to support individuals in Chapter 3, and organizations in Chapter 4. In Chapter 5, I covered monetization for third parties.

In the next three chapters, we will delve into the making of Cognitive Things. Chapter 6, looks at how Cognitive Things provide intelligent observations. Chapter 7, discusses their shared learning, and in Chapter 8, shows how they are designed, maintained and secured.

6

Intelligent Observations

6.1 Introduction

Herb Simon (Nobel Prize winner for Economics 1978) was a University professor at Carnegie Mellon and is considered to be the founding father of several of today's important academic disciplines, including artificial intelligence, information processing, decision-making, and problem-solving. I was fortunate to have him as an advisor for my Ph.D. committee, and had a number of futuristic discussions with him in the mid 1980s. He had an office full of books with an adjacent library. We used to sit in his library where he used to pile up many books on the table as well. One day, he dug in deep into his collection and showed me a set of Chinese sketches of social scenes and asked me to identify what a group of people was doing in each sketch. In a relatively short time, I found the first picture represented a village celebration while the second one represented a classroom. In his classic expression, Herb closed his eyes and paused for a minute before asking me the next question—"How will you teach a machine to recognize these social contexts?" In a matter of seconds, I had summarized a large number of pixels into a set of labels. Thirty years later, I can easily recall those labels, though the original pictures carried a lot more data, which my simple mind ignored.

IoTs are already generating extremely high volumes of event data. Most of this data is thrown away without any use, because it is bigger than any big data we have seen in the past. Over the last couple of years, I have visited a number of brave organizations, each of which had visions of collecting all the data and placing it into Hadoop big data lakes so they could analyze it. Unfortunately, once the data reaches the data lake, machines are limited in

A. Sathi, *Cognitive (Internet of) Things*,
DOI 10.1057/978-1-137-59466-2_6

what they can find.[1] Cognitive systems must observe, filter, and recognize like the way humans observe and summarize these observations to create meaningful patterns. The incoming sensor data must be filtered as close to the source as possible. Telecom organizations were the first ones to realize this problem. A typical view of a Facebook page on a tablet could easily generate hundreds of events in the network. These events can be trapped using network probes and sent over to a big data lake for analysis. However, the raw events may not be very meaningful and far too large in volume. A typical fifty million-subscriber network may create as many as 1–2 terabits per second of event data. In a typical busy morning, this may amount to a petabyte of data. How do we observe and find needles in the haystack in such a large volume of data?

In this chapter, I will describe ways in which Cognitive Things are beginning to observe intelligently and are able to abstract their observations. Like humans, machines are beginning to deal with many sensory inputs—vision, sound, taste, smell, and touch. The human mind is extremely good at dealing with these sensory inputs and extracting useful information for recall or reasoning. While Herb showed me those sketches for no more than 10 seconds about 30 years ago, I still have a vivid recollection of the room, the sketches, his face, and posture. Why do I not remember anything else about that day? While I do not remember the color of the shirt I wore that day, my wife can recall every piece of attire that she wore for every event for the last couple of decades. We know how to filter and store facts we care about.

In the automation era, machines improved their ability to ingest structured data and organize it into meaningful and governed data. As we enter the cognitive era, we must reuse the learning from structured data ingestion, such as data quality, governance, and veracity. In addition, we must introduce new cognitive functions like filter, focus, and chunking. Machines should be able to sense what is meaningful about hundreds of events associated with a Facebook page access and which ones to record and report.

6.2 Sense

In Chapter 2, I described many cognitive devices using a variety of sensors. The sensors are fairly good ways to take raw observations and convert them from an analog event to a digital encoding. For example, in the case of a Fitbit, the raw information is a step taken by an individual. The step could

[1] Arvind Sathi, "Big Data Analytics: Disruptive Technologies for Changing the Game", MC Press, Nov 2012, ISBN: 1583473807, https://www.amazon.com/Big-Data-Analytics-Disruptive-Technologies-ebook/dp/B00A2FOP04/

be large or small. Most tracking devices have a way of identifying events out of analog information they receive, and are able to summarize all that analog data into a digital event—in this case a step.

Sensors recognize and transmit their data using the format chosen by their manufacturers. As the number of devices, manufacturers, and technologies multiply, so do the formats associated with their data. For example, in the case of wireless telecom organizations, each vendor provides their data for 2G, 3G, and LTE networks. For the same manufacturers, the format may differ with each technology, and then there are differences across vendors. As these raw events are collected and observed using smart probes, which can recognize and summarize the events, the probes must understand format differences and normalize the data across many formats into a common representation.

Most of this data is fairly boring and useless for analysis or actions. For example, take the example of my security cameras. As long as there is no activity, or changes in the image, there is nothing for the camera to report. Sensors are becoming very good at setting thresholds for reporting. They report only when there is something interesting to report. For simple thresholds, it is easy to set them as conditions, but how do we deal with complex situations and aggregates? For example, each time I access Facebook on my tablet, it generates a number of events. While I may not be as concerned about a single image download on my Facebook page, I certainly would be concerned if Facebook downloads suddenly became very slow, or if movie downloads required unusual buffering. It is possible to create a composite score for service quality, apply that scoring algorithm to all incoming events, and report just the score to the server. The raw data can now be discarded. As requirements change, fresh scoring algorithms can be sent to the Cognitive Thing, making it far more adaptive and dynamic in responding to a changing environment.

This area is rapidly evolving and is termed "edge analytics".[2] Edge analytics means more than aggregation and filtering. It ranges from simple calculations to estimating the parameters of a statistical model using statistical languages like R or applying a statistical model to the data stored locally.[3] Instead of sending raw event data to the central database, which would clog bandwidth and cause enormous computing burden in the central storage, the analytics processes are distributed to the edge. The configuration of the edge may include a gateway device, connecting to a number of sensors and having adequate computing power and energy source for distributed analytics. The software application is the heart of the gateway. The gateway software is responsible

[2] Bosch Security, "Video analytics at the edge alerts you when needed and helps you quickly retrieve the correct data", YouTube, May 27 2014, https://youtu.be/drlAFbrJWwY

[3] Michael Hummel, "Geo-Distributed Analytics: Edge Analytics in the Age of IOT", DZone IoT Zone, September 19, 2015, https://dzone.com/articles/geo-distributed-analytics-edge-analytics-in-the-ag

for collecting messages from the sensors and storing them appropriately until they can be pre-processed and sent to the data center. The gateway software decides if the data at a given stage of processing should be temporary, persistent, or kept in-memory.[4]

So, what does it mean to my home security system? Instead of having a camera which records every time some one shows up at the front door, I can provide my camera with description of friendly guests, and the security system can let them enter the house, keeping it safe from unwanted intruders. As I acquire new friends or would like to "unfriend" someone, I could update the edge analytics scoring function. This security system would be better than my trained dog at guarding the house, as it takes a while how to teach a dog to "unfriend" someone. The sense function would work with minimal load on the broadband network as it uses a combination of sensors and edge analytics to do a fair amount of processing locally and sends only small amounts of filtered information for further analysis and processing.

6.3 Observe

Sensing was associated with taking raw sensor data and using edge analysis in a cognitive way to record and report significant events. Observations take our Cognitive Things to the next level of intelligence—making sense out of data over a time period and using a combination of distributed data ingestion and edge computing to reduce the bandwidth and storage requirements for the central server.

In Chapter 4, I introduced the work from IBM Research in the city of Nairobi using a set of smartphones associated with waste collection trucks to locate road conditions. Let me use a variation to that case study to illustrate the power of observations. Smartphones collect the location data using GPS and can report any changes in velocity of the moving vehicle. The waste collection trucks can be configured with additional sensors, such as cameras to take pictures. As the trucks move through the city, they can become mobile observers for a number of city projects, and can provide traffic as well as road conditions. The data is very useful and yet very repetitive. If all the camera and smartphone data is dumped to the central server, it may clog the server storage with useless raw data not worth observing. A smart observer, however,

---

[4] Henryk Konsek, "IoT Gateways and Architecture", The DZone Guide to the Internet of Things, https://dzone.com/guides/internet-of-things-1

would be looking for repeated instance of interesting events and reporting those events only. This may include recording a picture of the road and sending Latitude, Longitude, as well as the picture, any time the truck changes its velocity beyond a threshold. The analyst may decide to focus the data collection system only on those situations where the truck is slowed by adverse road conditions, and may be interested in quizzing the phone or camera for additional information to scertain the reason for slowdown.

Smart filters are able to make a series of observations based on filtering criteria. They may locally store a lot more data, and are able to provide additional data against a query. In the case of waste collection trucks making observations, the information collection can be set to provide minimal information in the first round to get a general observation of the city. As road conditions are identified, the camera can be activated in each repeat visit to the same street, especially if the smartphone observes the same speed reduction in each repeat occurrence. In this case, the central analytics server is working at the aggregate level and is using the smartphone and camera associated with the waste collection trucks to make focused observations, going into a lot more detail when there is a need to observe more detail.

Let us look at another example, concerning filter and focused observations from Counter Fraud Management. A fraud detection and management team would like to isolate methods fraudsters use to steal subscriptions and conduct illegal transactions. A very small percentage of transactions are fraudulent. However, according to the Communications Fraud Control Association (CFCA), they add up to $5.22 billion in annual losses.[5] Fraudsters keep finding new loopholes in telecommunications processes and systems to invent new ways of defrauding people. How can a fraud management organization make agile changes to its fraud detection and prevention software to effectively find and deactivate fraudulent subscriptions before they cause excessive revenue leaks?

This is a good application for intelligent observations using smart filters and focus. Smart probes associated with wireless devices can look for fraudulent conditions exhibited by a phone, such as geographically inconsistent locations. While most of wireless subscribers travel with their smartphones, they use typical modes of transportation to move from location A to location B. If a fraudster steals the identity of a phone and misuses the identity, it is very likely he will do so from a new location. Intelligent observations from smartphones can report sudden location changes, especially, if they

[5] "2013 Global Fraud Loss Survey," Communications Fraud Control Association. Note: registration is required to obtain a copy of the survey report. http://www.cfca.org/fraudlosssurvey/

are from a different location unreachable from the previous location via a normal transport mechanism. Once a phone is suspected to be in fraudster's reach, the central server can instruct detailed observations on the suspected phones, "focusing" on all usage, and making detailed observations on how the suspected phone is being used.[6]

6.4 Listen

Despite other ways of connecting with customers—such as smartphone apps, web page, and chat services—call centers still deal with the bulk of customers, ranging from sales to service requests. Call center agents are trained for several weeks to learn how to respond to these calls. Unfortunately, most call centers also experience high attrition rate leading to enormous training time wasted due to short tenure. Can we use a cognitive agent to listen to every conversation and, hopefully, start participating in the response, providing ways to augment, assist or replace the traditional call center agent? It makes sense to talk directly to the refrigerator or the television to complain about its problem, rather than a call center representative somewhere in another part of the world! The device may have additional information to confirm or challenge your questions, and may also debug itself during the conversation. To do all this, my cognitive device must learn to listen to my problems. How would my Cognitive Thing converse with me?

Speech analysis is a well-researched area, and is now reaching new levels of maturity. The original breakthrough program was created at the Carnegie Mellon University and initiated an entire field of research. The Hearsay-II problem-solving framework reconstructs an intention from hypothetical interpretations formulated at various levels of abstraction. The Hearsay-II system comprises problem-solving components to generate and evaluate speech hypotheses, and a focus-of-control mechanism to identify potential actions of greatest value.[7] The original system was limited to 1000-word vocabulary, and made correct interpretations for 90 percent of the test sentences. Today's systems are aspiring to deal with open speech in a call center, which, in addition to the words in a general dictionary, may use additional words specific to a particular organization. For example, the word "U-Verse"

---

[6] IBM Analytics, "IBM Counter Fraud Management for Telecommunications", YouTube, Mar 3 2015, https://youtu.be/vhl-JnLU6YU

[7] Erman, Hayes-Roth, Lesser and Reddy, "Hearsay-II Speech-Understanding System: Integrating Knowledge to Resolve Uncertainty", Blackboard Systems, Editors: T. Morgan, R. Engelmore, Addison-Wesley, pages 31–86, 1988, http://mas.cs.umass.edu/Documents/Erman_Hearsay80.pdf

has specific meaning to AT&T and its customers. When a customer starts with the sentence, "My U-Verse connection is not working", both the customer as well as the call center agent fully understand the meaning of this sentence.

If we were able to assemble all the words and understand their meaning, can we get the intent of a sentence? Consider the following sentences:

"I cannot login into my account?"
"My access to your website is locked?"
"What is my password?
"I have forgotten my password"

In all cases, most probably the user needs to reset password. In the first case, the question could be interpreted in other ways as well. May be the user cannot locate the login screen, or it may also be possible that he has forgotten his user ID. As the Cognitive Thing analyzes the speech utterances and formulates the sentence, how would it decide if the password needs to be reset, and respond with "Would you like me to reset your password?"

Natural Language Processing (NLP) is the area of artificial intelligence that examines human language and the different techniques to analyze, systematically process, and understand language. Natural language has an inner structure that is not explicit, and part of the task of automatic processing is to organize the language.

The process of analyzing language can vary greatly from language to language, as language incorporates the world vision and common sense of their speakers. Some languages share common roots and evolved influencing each other (for example, Spanish and Portuguese), while others can be independent of each other (consider Arabic, Russian, and Japanese).

Text analytics or text mining refers to the tasks involved in analyzing unstructured text and extracting or generating structured data, and performing some level of interpretation on that data.

There are many ways of collecting intent from a sentence, and each approach is pertinent to certain use cases. In the simplest approach, a key word search can match words used in a sentence to probable intent. For example, a number of traditional Interactive Voice Response (IVR) units graduate from a multi-layered menu to free-form questions and key words to guess the intent. The response can be in the form of a confirmation or disambiguation to match the most probable intent. Once the match is made the Cognitive Thing can respond by executing the stated action. The second approach uses statistical methods in matching sentences against a library of sentences. Given a large training set, the library provides many ways in which the same intent can be expressed. As long as the sentence ingested matches one of the sentences in the library, the Cognitive Thing can confirm, disambiguate, and respond. The comprehensive approach involves parsing the sentence into its components as tuples, such as subject–verb–object and understanding the intent by matching these tuples to known intents. This approach facilitates contextual interpretation, as a pronoun in a subsequent statement may refer to a noun in the previous statement. The sentence can be ambiguous and may be parsed in different ways depending on the context of the sentence.[8]

6.5 Crawl

In addition to speech, unstructured electronically readable text is rapidly becoming an important input for Cognitive Things. This may include social media, instruction manuals published by manufacturers, tutorials from third parties, complaints with government and related regulatory organizations, internal office emails, product documentations, and many more. The text is typically encoded in the format specified by the document management system, and may also carry illustrations, videos, tables, or other explanatory additions. There may be several purposes for ingesting and crawling the available documentation in order to:

- Search and bookmark relevant pages in support manuals and tutorials, so that the information can be made readily available, when needed.
- Extract relevant information so that it can be used as a summary in a standalone manner.
- Reconstruct a recipe or a tutorial from available documentation.

Unstructured data is the data that doesn't have an inner descriptive structure or definition appropriate for the intended task. When describing structure, data modelers refer to a type of organization present in metadata that accompanies the data, such as the column definition for a table in a relational database. In such cases, adding unstructured data to big data analytics might begin with the task to align the data to its implicit structure. In most cases, this processing is done to be able to aggregate, report, and act on the information inside the unstructured data.

[8] Dustin Wells, "IBM Watson Analytics—Natural Language Processing", YouTube, Jan 28 2015, https://youtu.be/tu5v-gu_5pY

A set of data can be considered structured or unstructured, depending on which task you intend to use it for. In its raw form, unstructured text looks more like a large collection of characters, text, numbers, and symbols to the analytic system, with a certain degree of organization. This data requires further transformation to be useful for analytic purposes.

Therefore, you also can interpret the term *unstructured* as a degree of organization of the data relative to a task to be accomplished. After this data is aligned to an analyzable structure, the data can be treated as structured data. For example, a binary file that contains an image can be considered structured data for the tasks of use and display by digital imaging software. At the same time, the same file can be considered unstructured data in the context of image outline object recognition.

6.6 Visual Recognition

At the beginning of this chapter, I introduced a visual recognition task given to me by Professor Simon. Human visual recognition is extremely good at scanning an image and converting it into a set of familiar terms. Whether a child is shown a photograph of a cat or its cartoon caricature, they are able to recognize it as a cat and differentiate from other animals. As we see a social scene, we can immediately guess the group activity these individuals are doing, whether sitting in a classroom, buying and selling in a shop, or dancing together.

Is it possible to teach a computer to do the same? This is a vital capability for Cognitive Things. As driverless cars proceed along the streets, they must recognize and react to a small child crossing the street. As a robot watches someone initiating a high-five, they must respond to the high-five. As the digital assistant records an accident, he must be able to identify a damaged car as well as the model of the car. In each of these cases, there is a visual recognition software, which ingests the visual information available from the cameras, compares the visual information to known images, and uses the comparison to identify each object, classify the collection of objects, and extract meaningful information to decide what that image represents.

The vision may involve color perception, for example, as a driverless car differentiates green from red at a set of traffic lights. The cognitive agent may need to have depth perception, involving some of the three dimensional objects and an understanding of their size and shape. The vision may involve feature extraction, thereby understanding and articulating differences between a smiley face and a frown.

Using statistical processes, cognitive systems can be trained in much the same way as humans to recognize and classify objects.[9] However, the training set must be adjusted based on the type of recognition task. It is a lot easier to differentiate a car from a motorcycle, much harder to differentiate between 2008 and 2010 versions of a car model. Statistical processes also assign probability of recognition, thereby quantifying the confidence level in the results obtained.

6.7 Identity Resolution

In a typical work environment, cognitive tasks require a speedy and yet careful resolution of identity to decide recognize an individual, a situation, or a resource. The source information may contain noise or poor input quality. The information may point to more than one individual and further investigation may be needed to precisely identify an individual. It may also include detection of deliberate deception or identity theft. Cognitive tasks often use background information and other data sources to eliminate candidates, resolve identity differences, and select alternatives using the context of the task. Let me use an example from media audience analytics to describe use of contextual information to resolve identities.

Let us go back to the cognitive television and the need for identifying the individual watching the television. Television audience data can carry some inherent ambiguities. While set-top box data is associated with a television set in a room, often it is hard to ascertain who was watching the television. The viewer might be an adult male or female, a child, or the family dog that is watching a television that was left on. However, a non-linear viewing on a mobile device can be precisely connected with an individual. In addition, robot televisions can test the programming, and might need to be removed from the audience data. With entity analytics, you can assign probability for these diverse sets of views and identities by using the available data. For example, the programming can be used to guess the gender and age of the viewer. Also, the non-linear viewing can be used to establish historical viewing patterns and applied to related set-top boxes. In addition to the viewing data, there may be additional background information available, such as on social media. People in social media reports, intentionally or not, frequently provide information about their viewing habits, preferences, and opinions on the latest episode of their series, and what they recommend that other people watch.

[9] IBM Watson, "IBM Watson Visual Recognition", YouTube, Dec 4 2015, https://youtu.be/n3_oGnXkMAE

The task of entity analytics is to create an individual profile of a person, to accumulate clues incrementally, and to resolve possible conflicts between conflicting information.

Entity resolution is the task of:

- Mapping and aggregating profiles that belong to the same person across the system.
- Removing duplicate profiles for the same entity.
- Accumulating context that can help resolve this situation in the future if it's not solvable right now.

Customer profiles have been a subject of focus for decades. Customer relationship management (CRM) tools were the first to offer an integrated customer database, one that would unite individuals and their family and maintain a household hierarchy. As organizations merged and departmental information technology (IT) investments were centralized, they found that each organization had a different view of the customer. To make the situation more confusing, all of these organizations used the same words—customer, product, address, order, and so forth to mean different terms. The central technical capability in any master data management (MDM) is its ability to match identities across diverse data sources. How do we integrate unstructured data sources with the matching capabilities of the MDM solution? Most MDM solutions offer matching capabilities for structured data. MDM software matches customers and creates new IDs that combine customer data from a variety of sources. These solutions are also providing significant capabilities for using customer hierarchies to normalize data across systems. However, in most cases, the format for the data is known, and the content is primarily structured. What is the cognitive resolution process where the data formats, definitions, and structures may vary widely?

The difference between a conceptual entity model and a typical structured model is that the structured model is focused on defining a structure for a database in a formal language. With unstructured data coming from a variety of sources, the conceptual entity model would rely on an ontology that assembles commonly known definitions, and attaches them to content or structure of a subject matter. Entity resolution is the process of combining multiple data sources to disambiguate real world entities from different sources to one single conceptual entity. The additional entity assertions may be probabilistic in nature. That is, the interpretation may carry a level of confidence represented as a percentage. Identity resolution processes may change those probabilities as it finds new correlations among seemingly uncorrelated activities.

- Deterministic methods rely on algorithms that are created specifically to match entities, considering the special characteristics of the data sets being analyzed.
- Probabilistic methods rely on statistical methods to determine the likelihood of a linkage between a set of entries in the real world to one single conceptual entity.

Consider the television audience example to illustrate the identity resolution process. By combining social media data, smart tablet viewership information, location information, and set-top-box viewing, I can formulate the digital identity of each individual in the family using a fuzzy probabilistic model. An ontology can be built, which establishes members of the family, media viewing devices, events, and programming. Different data sources can now be mapped to this ontology to understand how they contribute to the understanding of the media viewing. Based on historical data, a digital signature can be established for each household member, and recent information can be used to override any past trends.

6.8 Chapter Summary

In this chapter, I described a number of technologies involved in ingestion of unstructured and big data sources. Cognitive Things are facing the same dilemma individuals face—the data tsunami. For efficient information processing, Cognitive Things are learning to focus, filter, and select data based on downstream needs. *Analytics at the edge* uses filtering and focusing criteria from downstream applications to significantly reduce and abstract the collected data before sending it to downstream systems. The result is a reduction in storage and network throughput requirement, without any loss of information content.

In dealing with individuals and organizations, the input information may contain a fair amount of *unstructured* voice, text or visual information. This chapter provided an overview of techniques for speech analysis, natural language processing and visual analysis for these sources. These techniques convert unstructured information into a variety of data structures suited for downstream processing. Cognitive learning techniques are being used for converting free-form speech to text, for natural language processing of unstructured text to identify intent, and for visual recognition of objects in a visual. Extensive training of Cognitive Things enables them to make finer classifications, whether in listening to foreign-accented English, making sense out of misspelled words, grammatically incorrect sentences, or for identifying a red light on a foggy day.

Cognitive Things also organize data sources and isolate identities providing the data. *Identity resolution* involves attributing a data to a source, by using context information, and through rules and criteria-driven elimination of alternatives. In this way, they can identify the audience for a television or a web user on a shared desktop.

In the next chapter, we will look at information processing techniques for organizing, analyzing or deciding on this data. In Chapter 8, we see how Cognitive Things get efficiently installed, trained and maintained, thereby expanding further on the training aspects briefly covered in the current chapter.

7

Organization of Knowledge and Problem-Solving

7.1 Introduction

Cognitive scientists have conducted many studies to compare decision-making across novices and experts.[1] Each human has the same short-term memory constraint. We remember five to seven things. Try remembering a 10-digit number and your head will hurt after 30 seconds. When I remember a 10-digit US number, I cheat by converting the three-digit area code into its city equivalent (thereby converting three items into one) and then try to decipher a pattern for the other seven digits. Professor Herb Simon has termed this as "chunking". Once chunked, experts store a much larger number of chunks in their long-term memory. In his studies of chess, he found that experts were far superior at reconstructing chess game positions as compared to novices. However, when he used random chess positions, experts did not see any chunks and were no better than the novices. Comparison of a computer simulation with a human experiment supports the usual estimate that chess masters store some 50,000 chunks in their (long-term) memory.[2]

How do experts build such a large collection of chunks? Try eavesdropping into a conversation between two medical doctors, and you can spot these chunks in their conversation. Unlike other living beings, human learning does not come from genetic encoding, but from observations, teaching by experts,

[1] William C Chase and Herbert Simon, "Perception in Chess", Cognitive Psychology 4, 55–81, 1973, http://matt.colorado.edu/teaching/highcog/fall8/cs73.pdf

[2] Fernand Gobet and Herbert Simon, "Recall of Random and Distorted Chess Positions: Implications for the Theory of Expertise", Memory & Cognition, 1996 24 (4), 493–503, http://link.springer.com/article/10.3758/BF03200937

A. Sathi, *Cognitive (Internet of) Things*,
DOI 10.1057/978-1-137-59466-2_7

reading books, watching videos, working in the field, or any other form of input we may be able to gather. We can connect the dots, try new combinations, or follow our experientially learned instincts. In the case of medical experts, the chunks are gradually created over years of education, residency, fellowship, reading medical journals, and working with patients jointly with other experts. By working in an area for a number of years, these experts have created chunks in their cognitive memory, and are able to connect a set of associated facts together. In any field, years of training and experience leads to a well-organized representation of knowledge. Humans are amazing cognitive machines as they effectively use their very limited short-term memory of 5–7 items, while connecting the dots and recalling chunks from their long-term memory as needed for a specific problem-solving session. As new means for knowledge capture, storage, and retrieval become available, experts are increasing their reliance on recorded or documented knowledge and field experience to augment their long-term memory. Sharing is an important aspect of knowledge; for example, using books, journals, conferences, blogs, and consultations with other experts.

As the digital revolution makes large bodies of unstructured information available with Internet connectivity, it is extremely hard for an expert to keep up with changes. Is there a way to index all the available sources, and access context-specific information during problem-solving? Knowledge-based reasoning brings the next level of problem solving, as now this knowledge can be utilized in a more active form. Stored knowledge can be applied in many ways—either automated, or in a decision-supported environment for others to follow. The shared knowledge can be applied via Cognitive Thing, such as appliance (trouble shooting conversation), a driverless car (road navigation), or a chatbot (digital assistant). The experts contribute to the organizational knowledge base, growing the collaborative expertise and sharing with other junior employees in an organization as well as with the Cognitive Things.

This chapter explores the concept of knowledge representation and problem-solving. The first task is to identify how problem space can be best represented and how relevant information can be organized in support of problem-solving. The structures for knowledge representation must provide easy mechanisms for its use during problem solving, as well as its maintenance. Second, we will look at how knowledge can be extracted from these unstructured sources. Most of the knowledge sources are in the form of occasionally organized unstructured text, for example vendor manuals, government regulations, company best practices, individual cheat sheets, and so on. Digital data sources have created a large repository of semi-structured data, including web logs, purchase records, usage logs, locations, and so forth. The information contained in these knowledge-sources must be transformed into the target structures for knowledge representation.

Thirdly, I will describe how raw data can be converted into meaningful inferences to be accessed by problem-solving engines. Next, a set of examples will illustrate how reasoning engines can be assembled in different ways for a variety of use cases for individuals as well as organizations. Problem-solving can be as simple as selection across alternatives, or as complex as design through discovery and learning. Cognitive Things are accelerating our pace of decision-making. Driverless cars and home security systems cannot work in batch processing mode. I have included a section on real-time decision processing, highlighting how a Cognitive Thing uses historical and real-time inputs, and accumulated knowledge to combine predictive and prescriptive forms of decision-making.

7.2 Organizing Solution Space

The world is full of knowledge sources and a decision maker could be lost without its organization and prioritization. The first task anyone does when they start solving a problem is to organize the solution space by bringing relevant facts, constraints, goals, and alternatives to the forefront, as a consideration for the solution definition and decision-making. Take the simple task of organizing a meal. I have a *goal* to eat healthy because I am overweight. However, I do want to make the food delicious, which I would like to place as a *preference*, thereby choosing something that is healthy and yet tasty. I have another constraint in that my refrigerator and pantry have certain pre-stocked items. What are my *alternatives*? I can cook from what I have in the refrigerator and pantry. I can go to the grocery store to get additional items, but that would add to the meal preparation time. Alternatively, I can get a pre-cooked meal from the grocery store, or order from a meal delivery service. So, what is *out-of-the-box thinking*? After working with my goals and constraints and agonizing over the choices, I decide to go to my favorite restaurant for a delicious healthy meal and ask three friends to accompany me for delightful social company. In this option, I added new goals, got rid of constraints, and ended up with a completely new set of alternatives!

So far the focus was on my own goals, constraints and alternatives, how do I now add my friends' goals, constraints and alternatives. Anyone working for a large organization is fully aware of the long-drawn-out meetings, in which several colleagues or business partners hash out common solutions that span across shared goals or constraints. Extending the example above, if the same decision-making were to occur between two friends living in separate households but planning a meal together, the goals, constraints and alternatives include many more permutations and combinations. I may be interested in eating healthy, but my friend may not be, and now we need to find a solution that satisfies both goals. I may have half the ingredients in my refrigerator and the decision on where to cook may depend on who has the most ingredients and how easy it is to move the content from one house to another. While my friend and I may use a fairly unproductive and possibly stressful 30-minute call or 30+ instant messages to explore the solution space, our digital assistants may be able to collectively work for us and organize our solutions space in a couple of seconds.

Would we trust the assistants in gathering the solution space? This is assuming our assistants are capable of gathering all the data from the refrigerator and the pantry without missing a couple of items in side storage. Where would they go to gather our goals? What if the goals as listed on my smartphone are from last year, and I have not yet updated my goals? Data quality and governance will play as much a role here as they would in a typical Customer Relationship Management (CRM) application. If the assistants came up with poor quality data or obsolete constraints, it would be hard for us to accept their solution. Moreover, if it takes as much time to update constraints and goals manually, my friend and I will ignore the technical route and will resort back to the 30-minute-long call or serial texting.

In an organizational setting where collaborative decisions are a way of life and employee productivity can differentiate leaders from laggards, Cognitive Things may provide productivity enhancement and replace some of the human communication. The Information Technology (IT) department would like to find the best wireless communications provider for the corporation. Instead of sending a questionnaire to all the employees, the IT department decides to install an app on their smartphones to measure service quality. While only 2 % disgruntled employees would have responded to the questionnaire, the app would now provide service quality information for all the employees carrying wireless phone service, and now the IT organization has a far better organization of the solution space, even with geographical and time-of-day distributions to conduct a better negotiation with their wireless service provider. Alternatively, the wireless service provider may offer this as a value added service to the corporations.

Let me now take a problem where organizational knowledge is available, even if it is not well organized. For example, take the case of an industrial or telecommunications organization using a vendor-supplied equipment. The equipment is used in a mission-critical operation, where an outage may result in unavoidable business operation interruptions. It is in the best interests of the vendor and the consuming organization to keep it well maintained. In most cases, vendors provide a large number of sensors that monitor the equipment. The vendor's engineering department creates a series of troubleshooting procedures at the time of engineering design, enumerating malfunctions, test and repair procedures. The vendor may have provided this manual either online or in the form of a PDF document, which can be downloaded, or printed. As the equipment breaks down, the field service engineer discovers new malfunctions, test or repair procedures, not previously envisaged by the engineering department. These guides can now be organized in the form of a *Knowledge Graph*, which represents knowledge using a graph (a set of nodes and arcs). When the junior technician shows up at the next field visit, he traverses through the Knowledge Graph to diagnose and repair the equipment. After testing the knowledge graph in field several times, the equipment can use its sensor data, redundant components and software downloads, to self-repair by using the Knowledge Graphs originally created by the engineering department and updated through each and every field service visit.

In my last visit to the IBM's Almaden Research Lab, I was fortunate to meet with Anshu Jain, who has invested many years in developing Knowledge Graphs. His research work has resulted in a tool named Drishti, which has been applied to defects in Information Systems.[3] IT services is a complex area for troubleshooting. Anshu studied the work performed by quality analysts in a large IT services organization, where the evidence collected using trouble tickets was in a high-volume environment, and simple observations and spreadsheet tracking was unable to scale to the large number of failures and related events. The IBM Research team working on this problem developed an intricate combination of statistical analysis and text processing to extract Knowledge Graphs and apply them for finding and resolving defects.

In the research dating back to mid 1980s, Mark Fox, Mike Greenberg and I worked on an organization of Knowledge Graph, where each layer would inherit its structural properties from the layer below. In the research across many industries and domains, we found a common layer, which was industry- and domain-independent and provided foundation for the representation of organizational knowledge. The layers proposed in our 1985 IEEE article are shown in Table 7.1. In any field, expertise is layered.[4] A typical medical student learns a fair amount of biology and chemistry before getting into

[3] Anshu Jain, Srikanth Tamilselvam, Bikram Sengupta, Krishna Kummamuru, "Reducing Defects in IT Service Delivery", Service Operations, Logistics, and Informatics (SOLI), 2013 IEEE International Conference on, 2013, pp. 13–18, http://researcher.watson.ibm.com/researcher/files/in-srikanth.tamilselvam/SOLI_ReduceDefects.pdf

[4] Arvind Sathi, Mark Fox, Michael Greenberg, "Representation of Activity Knowledge for Project Management", IEEE Transactions on Pattern Analysis and Machine Intelligence", 10/1985; PAMI-7(5):531–552. DOI: 10.1109/TPAMI.1985.4767701. Also available as Carnegie Mellon University Technical Report, http://repository.cmu.edu/cgi/viewcontent.cgi?article=1565&context=robotics

Table 7.1 Knowledge Representation Layers

Layer	Definition and examples
Layer 5—Domain Layer	Represents concepts, words and expressions specific to a domain of application, such as industrial equipment, events, failure points, repair procedures, etc.
Layer 4—Semantic Layer	Represents common primitives such as time, activity, state, ownership, currency, etc.
Layer 3—Epistemological Layer	Represents flow of information through inheritance, such as classification, abstraction, aggregation, instantiation
Layer 2—Logical Layer	Represents concepts as a collection of assertions, for example using subject–verb–object to represent a fact
Layer 1—Implementation Layer	Represents primitives for machine interpretation of the concepts and assertions, such as nodes and arcs

advanced medical science topics. A typical electrical engineer learns physics and mathematics before getting into electrical engineering topics. The layering shown could be used as a structure for Knowledge Graph. In each layer, the concepts can be discovered automatically using statistical processes, or handcrafted by experts. The layers provide discipline and underlying framework for inheritance.

Knowledge Graphs are evolving in many areas. Medicine is probably the most important area for organizing medical expertise using the Knowledge Graph. Google has said "one in 20 Google searches is for health-related information." Yet, the information available in search results can be incomplete or untrustworthy, though there are many credible sources, as well. To improve the quality of health-related search content, Google is introducing structured and curated health information. Google has tapped doctors, medical illustrators and the Mayo Clinic to develop in-depth information for more than 400 health and medical conditions.[5]

Additionally, many tools are providing ways of extracting Knowledge Graphs and associated indices into document sources. In the next section, we will explore how text analytics is used for natural language understanding to create this organization of knowledge.

[5] Greg Sterling, "Google Introduces Rich Medical Content Into Knowledge Graph", Search Engine Land, February 10, 2015, http://searchengineland.com/google-introduces-rich-medical-content-knowledge-graph-214559

7.3 Text Analysis

Most of the knowledge sources use unstructured text to store the knowledge. Eliciting meaningful information from unstructured knowledge sources is a multi-step process. In this section, I have provided a broad outline of the technical steps involved in preparing the sources, extracting the data, understanding the meaning, and extracting facts. Often, binary files such as voice data or even image PDFs exist, which require pre-processing to extract the text to a form that can be further processed.

Normalization Text normalization is a low-level task that consists of transforming text into a single canonical form. As different languages have different machine representations, text normalization also defines encoding, format, and character set. For example, a .doc file created by Microsoft Word has many more control characters than a .txt file. In order to further process, the normalization process would strip off all these control characters from the .doc file. The platform needs to support multiple languages and multiple encoding formats, while it transforms inputs to a single standard to be used in other higher-level tasks.

Language Identification The next step to analyze any document is to determine the contents of the document, and in which language the document is written in. Language identification is the process of determining the source language of any document. Working with real data, one frequently finds that conversations often include more than one language. This complexity complicates the tasks of determining the language used in a document.

Tokenization Tokenization is the task of separating the different constituents of a sentence, or tokens. In general, this task separates numbers, punctuation, and words. In some languages, this task also involves adding separation between words for those languages that do not use spaces between words. Tokenization is an extra step to simplify the problem of automatically processing written language. While most tokenizers are deterministic, they can include language-dependent heuristics.

Part of Speech Tagging Part of speech (POS) tagging is the task of assigning a label of a part of speech to each word from a sentence. This task is often complicated by the nature of the language. Some parts are implicit and some parts are ambiguous. The task of speech tagging is necessary to address other higher-level tasks, as POS is a clue that can be used to increase the accuracy of other types of analysis. For example, POS can help you determine sentiment by analyzing whether a word is used as an adjective instead of a noun.

Feature Extraction Feature extraction, in the context of pattern recognition, is a task that aims to search, find, and classify inputs according to pre-defined classes to be extracted. For example, a US phone number can be recognized by a set of three digits in parenthesis, three more digits, a dash, and then four digits.

Named Entity Recognition The task of Named Entity Recognition consists of identifying specific entities in a text. The predefined categories or groups of elements have specific semantic relations, which are relevant to the topic being analyzed. Often taxonomy in a knowledge area provides a list of entities to be discovered in the source document. Each term in the taxonomy can be represented as a named entity and searched using pre-defined patterns.

Stop Word Removal Stop word removal consists of the task of identifying words that might not add to the semantic content of a specific task, and removing them. For example, keyword search indexes will typically not index occurrences of common English articles such as "the".

The stop word concept is brittle, and a word that in some context might be considered a stop word, might not be considered a stop word in others. Furthermore, stop words normally have a purpose in most communications. While their removal helps with "bag-of-words" approaches, they can be useful or even necessary if the analysis task is different.

Sentiment Analysis Most of the business use cases require the performance of higher semantics on natural language to obtain the most value from social media data. Sentiment analysis has received much attention in recent years, as more companies analyze social interaction to extract customer preferences. Sentiment analysis is a brittle task, which means that what works for one subject is not directly applicable to other subjects. Most tasks today use aggregations of sentiment relative to different topics or products, but it is becoming important to link individual opinions to individual subjects.

Information Retrieval Information retrieval is the task of finding information in a collection of documents. The initial tasks that are associated with information retrieval, which predate the Internet, were indexing and retrieving information in collections of documents.

The information query can be posed as a question, as a search pattern, or many other forms. The answer to this search can be of different types: a direct answer, a collection of references, a list of people that might know the answer, or even a clarification request.

The approach and special characteristics of the answer and the query differentiate the tasks in this field. For example, if the query is formulated as a question, and the expected answer is a single input with the correct answer, the task is a question answering task.

Automatic Text Summarization Automatic summarization is the process of reducing a document to create a second document that is called the summary, which contains the most important messages of the original document.

Text Classification Text classification is the task of assigning a label to a document. This general approach can be used for multiple purposes. Social media and microblogging services have received much attention in the last few years. More importantly, text classification can be used to automatically group similar opinions that are based on different segmentations.

Spam filtering constitutes another use case of text classification. In the context of social media data, spam filtering has been recently defined as the removal of social interactions that are produced by scripts or automated bots. The discovery of spam in social media has been of great importance in recent years, as it negatively affects most of the use cases of analytics on social data.

Relationship Extraction Typically, relationship extraction involves discovering how two or more named entities relate to each other. By analyzing the context of the named entities, such relationships can be identified and extracted. For example, a technical blog may identify an expert as an employee of a company or belonging to a community of experts. In such cases the relationship between expert and his employer can be identified and extracted.

Social media services provide a subset of the individual characteristics or features for each individual, but the analysis of the interactions between them can provide much more information. For example, by analyzing the interactive posts in social media (users who comment or respond to other users), you can extract some network relationship between people. Analyzing the content of those interactions can determine the nature of the relationships between them.

This task can be divided in two subtasks. Initially, you need to create profiles of the entities to be analyzed.

You can then define which relationships you need to extract. For example, in the case of fraud analysis it is likely that you would want to analyze family relationships. In churn analysis, understanding social media relationships is relevant to determining influence in the network.

Question Answering The question answering task is a subtask of information retrieval, where the answer is expressed not as a collection of documents, but as a single point of data that contains directly the answer to the query.

This task is divided into two main subtasks. The initial part of the problem consists of understanding the question, as it is expressed in natural language. Part of the first subtask includes determining the type of data that answers the problem, and the topic that the question is using. The second is to find the appropriate information that answers the query.

7.4 Profile Enrichment

I will use my favorite example of a customer profile to illustrate how raw sensor data can be abstracted into a set of profile attributes about an object. Profile enrichments are useful in reasoning through a problem, as they reduce the amount of data to be analyzed by the problem-solving system. The enriched profile is typically extracted from transactional, demographic, conversations and usage data sources. These enrichments provide a richer insight to behavior, preferences, and usage patterns. The resulting customer profile that is built with a much larger data set generates thousands of micro-segments that can be used across various use cases in sales and marketing, customer care, revenue and risk management, and resource management. An organization can employ these profiles in business processes for fraud, customer care, or marketing.

Table 7.2 shows examples of data sets associated with each data dimension.

7.5 Automated Problem Solving

Much of the discussion so far in this chapter has been about problem-solving involving known components. Cognitive Things have improved information access for comparing alternatives, and have been used to carry out decision-making in an effective way. Problem-solving is not always that easy. The alter-

Table 7.2 Sample data dimensions and data set

Data dimension	Example data sets
Description	Age, income, gender, education, family size, home location, occupation, ethnicity, nationality, religion, federated ID, CRM ID, billing telephone number, Government ID, family unit, organizational hierarchy, social media IDs, email address
Interaction	Channel interactions, third-party interactions, contact preferences, alerts, billing history, payment history, subscriptions, privacy preferences, NPS survey
Behavior	Locations, usage, device failure patterns, fraudulent patterns, socio-temporal patterns, usage lifestyle, personality, fraud alert, care alert, lifestyles, activities, media viewership patterns, customer lifetime value, favorites, social network, social media discussions, work location
Attitudes	Sentiments/opinions, brand loyalty, usage experience, social leadership, attitudes, derived net promoter score

natives may not be always available or obvious. Can Cognitive Things create a new alternative from past experience?

Many route-planning apps are already doing that for the drivers. Using Waze or Google Maps, a driver can specify a starting point (often the current location) and an ending point. The route planners are getting increasingly sophisticated in dealing with real-time traffic information, often crowdsourced from other drivers. In creating these routes, the apps use complex algorithms that use a combination of mathematical programming and heuristics to find the routes in real-time.[6] In computing routes, these algorithms are finding new routes, learning from past routes taken by other drivers, and comparing routes to choose the best. In many cases, the app also shows the driver various routes and compares driving distance or time from point A to point B. What are the general characteristics of problem-solving and how do cognitive engines generate alternatives? Here are several ways in which algorithms create new solutions.

While route planning is a complex task, it is simple in its data sourcing. Using maps, traffic congestion data and the destination information, it generates a route from point A to point B. What happens if I now have many inputs, many algorithms, and need an orchestration mechanism to combine them? Take the example of a personal assistant. In order to do simple tasks like travel planning, a personal assistant needs access airline, hotel, and car rental

[6] Hannah Bast et al., "Route Planning in Transportation Networks", Technical Report, MSR-TR-2014–4, Microsoft, http://research.microsoft.com/pubs/207102/MSR-TR-2014-4.pdf

sites, extract information on alternatives, use the context of the overall trip to rank order options and make decisions.

Viv (see YouTube video referenced below[7]) breaks through those constraints by generating its own code on the fly, no programmers required. Take a complicated command like "Give me a flight to Dallas with a seat that Shaq could fit in." Viv will parse the sentence and then it will perform its best trick: automatically generating a quick, efficient program to link third-party sources of information together—say, Kayak, SeatGuru, and the NBA media guide—so it can identify available flights with lots of legroom. And it can do all of this in a fraction of a second.[8] Viv comes from the folks that created Siri and sold it to Apple. But, unlike Siri, Viv will be an open platform, which means it can be programmed to work with just about any app or service. For the most part, Siri only works with Apple's own apps, though there has been talk of a Siri API to open up to other iOS developers. But Viv won't be limited to a single platform. It could run on other phone operating systems or be integrated into other devices such as entertainment systems, home appliances, or perhaps even cars.[9]

A number of reasoning engines can be defined for alternative generation. The following discussion illustrates a couple of these reasoning engines. Most of the cognitive work today uses "Alternative Generation using Selection". Using the taxonomy of a subject area, a classification reasoning engine can select alternatives.

Alternative Generation Using Selection The simplest way a set of alternatives can be created is by going to a library of alternatives, and selecting those which meet certain criteria (for example eliminating those alternatives that do not meet the goals), and then comparing the chosen alternatives using more detailed evaluation criteria. Depending the role, the Cognitive Thing may either choose the best alternative, or present the results to a human for final decision-making. A good example of this is campaign execution, where a set of campaigns may apply to a micro-segment and the campaign management system may choose the campaign with the best probability of being accepted by the customer at the least cost.

---

[7] Dag Kittlaus via Tech Crunch, "The team behind Siri debuts its next-gen AI "Viv" at Disrupt NY 2016", YouTube, May 9, 2016, https://youtu.be/MI07aeZqeco

[8] Steven Levy, "Siri's Inventors are Building a Radical New AI That Does Anything You Ask", Wired, August 12, 2014, http://www.wired.com/2014/08/viv/

[9] Larry Magid, "Move Over Siri. Your Parents Just Gave Birth To Viv. And It's Much Better", Forbes, May 9, 2016, http://www.forbes.com/sites/larrymagid/2016/05/09/move-over-siri-your-parents-just-gave-birth-to-viv-and-its-much-better/#63b5136ea066

Alternative Generation Using Configuration In this approach, a library of components is available, and the algorithm chooses a set of components based on configuration constraints and user goals. The end result is a new alternative, which is made up of known components. The route-planning algorithms described above are good examples of this approach, as each algorithm uses a set of route segments and configures them for the overall route from point A to point B.

Alternative Generation Using Extrapolation In this approach, a new component may need to be created by modifying an existing component or the overall solution. Typically, a configured solution may be close to an accepted solution but could be improved by changing a parameter. By modifying the solution, it becomes more acceptable. For example, a shopper's assistant may find a product I may be looking for but may recommend a delay in purchase due to an upcoming promotion, which would reduce the price I would pay for the product.

Alternative Generation Using Discovery In this approach, a new alternative is discovered by searching for a solution. Typically, this may involve looking for how others have solved the problem. In a typical product support environment, a field technician may use past cases to look for a manifested problem in equipment and how others fixed the equipment.

7.6 Adaptive Real-Time Decision-Making

Effective targeted actions require analytics and process infrastructures for detecting, focusing, and designing an action. In many cases, the action must be taken in real time while a customer is interacting with a provider or walking past a point-of-interest location. The solution requires tight integration and performance management across a number of its components. The D4 framework offers an integrated solution for these targeted actions using four components that discover, detect, decide, and drive targeted actions (see Fig. 7.1).

An analyst uses analytics techniques to discover historical behavioral patterns. Through rigorous quantitative and qualitative analytics processes, the analyst may establish normal as well as abnormal behaviors. For example, a marketer working for the New Store retailer may *discover* a micro-segment that likes to collect Foursquare points for visiting grocery stores and also regularly shops at the Good Neighbor department store. In addition, the

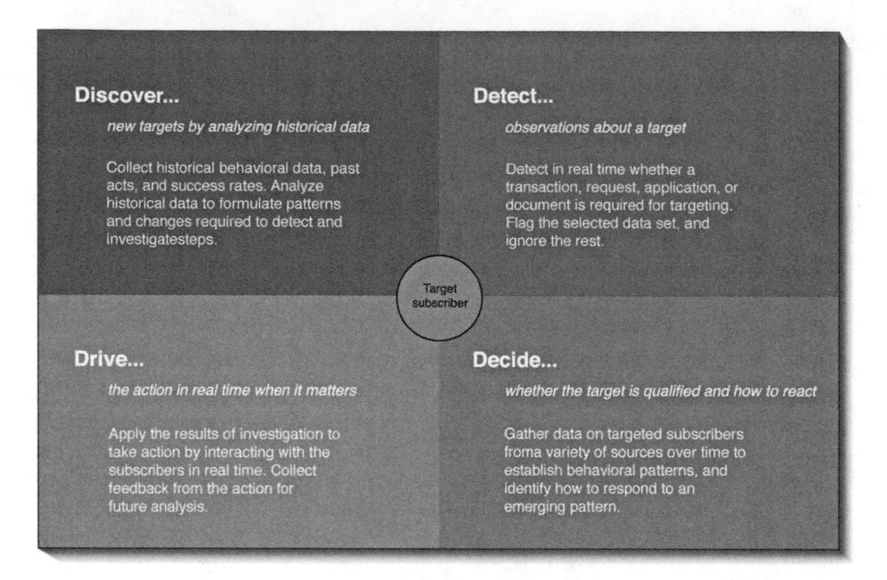

Fig. 7.1 Discover, detect, decide, and drive using advanced analytics platform

marketer observes Lisa, a customer who regularly visits the Good Neighbor department store every Thursday evening when she is not traveling out of town for work. The associated parameters and related thresholds are now passed on to the *detect* component, in which big data sources are scanned on a near-real-time basis. This component discovers when a collection of observed parameters matches a desired pattern. In the campaign example, this matchup could be the Thursday when Lisa is not traveling for work and should be offered an incentive to visit the New Store. In doing so, the real-time analytics system may ignore all of Lisa's neighbors, who do not match the specific pattern. The selected customers are then passed on to the *decide* component, in which a rule engine configures a specific action to handle a particular situation. In the case of the marketer campaign example, the marketer may use a variety of rules covering saturation—how many offers were made in the recent past—along with response likelihood based on past offer redemptions and predisposition to location-based offers. As a result, two members in a micro-segment may receive radically different offers to maximize the likelihood of redemption. The chosen decision is now conveyed to the *drive* component for the business process action. This process is a campaign management system that engages with the customer, delivers the action, and records the customer response to fine-tune the parameters for future offers. The architecture requires a tight integration across four divergent tools. The discover component is mainly conducted using statistical and machine learning tools. The detect component uses real-time analytics products, which can ingest large-scale data and filter it based on known criteria. The decide component uses rule engines that use a combination of real-time data and historical data to configure a specific action. And the drive component uses business process automation and workflow to drive the decision.

The solution requires an intricate combination of three underlying advanced skill sets: data engineering, data science, and application and domain knowledge. Data engineering is required for high-volume, high-velocity data. Given the level of automation, the entire decision-making cycle should be completed during a transaction. It requires a system that can react to changes in near-real time ranging from a couple of seconds to a couple of minutes, depending on the use case. At the same time, both discover and decide components represent large amounts of data to establish patterns and for historical analysis of a chosen focus area. Data engineering also needs to institute the appropriate data flows without choke points to help facilitate smooth interworking across these components. Data science is required to handle patterns that look for exceptions. Working with averages is no longer sufficient. In many use cases exceptions rule. Data scientists should establish ways to identify micro-patterns and methods of detecting and deciding to find and act on these micro-patterns. Application and domain knowledge is necessary because the pattern covers the entire gamut of the decision component. Someone who has a very good understanding of the use case and the associated industry domain knowledge can easily weave an intricate interplay across the elements. While an outsider may develop surface-level algorithms and rules for each component, a domain-savvy expert can identify an underlying domain model to help drive all four components.

So, where would Cognitive Things fit into this campaign management system? First, the Cognitive Things provide the much-needed data acquisition component, by collecting valuable usage and location data from consumers. Second, the Cognitive Things become the shopping assistants who can assist a shopper, while at the same time collecting and refining their recommendations based on incentives offered by marketers. The Cognitive Things can now provide easy access to these incentives while considering other features such as price, product quality, brand, feedback from other customers, and any other inputs available to the shopper assistant.

In the world of marketing campaigns, there are two competing forces at work. On one hand marketers must focus on consumer need and behavior and offer products which best meet consumer needs. On the other hand, the marketers have campaigns to promote new products, new ways of reaching different micro-segments, price reductions to get rid of excess inventory, and so on. The shopper assistant works collaboratively with the marketers to ingest the best available information from the marketer, but also pulling information from other sources, and doing comparisons across marketers.

The increase in volume and velocity of information for both buyers and sellers is resulting in an increased number of campaigns. Many of these campaigns can be executed in parallel on test markets using A–B testing where one part of the market is offered one campaign, and another part of the market is offered a competing campaign. The comparisons are done using machine learning, often without requiring manual interventions. Marketing programs can simultaneously execute and compare hundreds of campaigns and make changes based on environmental conditions, test results, and changing trends. External data, such as weather can be used as additional input to make appropriate changes to campaigns.

The real-time adaptive platform can be applied to many other use cases, each of which may require interactions with micro-segmented customers. For example, in the proactive care use case described in Chapter 4, the cognitive agents can be used for providing improved care to the customers.

7.7 Chapter Summary

This chapter explored the concept of knowledge organization and problem solving. We looked at the concept of chunking and what differentiates an expert from a novice in their organization of chunks in their cognitive thinking. I described how Cognitive Things are able to create similar chunks as well as use these chunks for problem-solving.

I also introduced the concept of Knowledge Graphs. Most of the knowledge sources are in the form of occasionally organized unstructured text, for example vendor manuals, government regulations, company best practices, individual cheat sheets, and so on. Knowledge Graphs place a structure to these unstructured knowledge sources, thereby facilitating connectivity across concepts for problem-solving.

Most of data sources for extracting Knowledge Graphs are in unstructured text with occasional tables and cross-links. Text analysis provides us with a toolbox for analyzing unstructured text and extracting meaningful structures, which can be organized into Knowledge Graphs. Text analysis reformats the data, so it can be parsed, and then uses a series of processes to understand their meaning and purpose.

Digital data sources have created a large repository of semi-structured data, including web logs, purchase records, usage logs, locations, and so forth.

I used customer profile as an example to show the use of historical data to create meaningful dimensions which describe behavior, preferences, or demographics of an individual.

Next, we looked at a set of examples to illustrate how problem-solving components can be assembled in different ways for a variety of use cases for individuals as well as organizations. Problem-solving can be as simple as selection across alternatives, or as complex as design through discovery and learning. I used a couple of examples to show rapid auto-generation of processing modules, using standard components, which can be configured together for a specific problem-solving.

Most Cognitive Things cannot work in batch processing mode. A section on real-time decision processing has been included to highlight how a Cognitive Thing uses historical and real-time inputs, and accumulated knowledge to combine predictive and prescriptive forms of decision-making. The adaptive decision-making uses supervised and unsupervised learning to improve on its decision-making and uses field testing to experiment with alternative decisions.

The task of creating a Cognitive Thing involves configuring data sources, structured knowledge, problem-solving techniques and decision execution. The next chapter discusses how Cognitive Things are installed and how they learn to organize and improve on their knowledge and problem-solving.

8

Installation, Training, Maintenance, Security, and Infrastructure

8.1 Introduction

This chapter is where the rubber hits the road. So, you were very excited to try out all these Cognitive Things and ended up increasing the size of your cognitive household with a number of new shiny objects, that think and empathize with you, but you are adding new mouths to feed (in your time and effort), and now your energy is spent configuring and maintaining these Cognitive Things. You may find yourselves searching for discounted batteries at Home Depot to keep providing juice for your Cognitive Things, or marveling over the large key chain you have amassed on your laptop to keep track of all the passwords. Did you gain or lose in time, convenience and productivity?

Remember Tamagotchi, the egg-shaped virtual pet that dies on its owner daily unless afforded generous love, care, and attention.[1] They were a wild success in 1990s, and we all loved them. The Tamagotchis showed a lot of emotional drama and attitude and needed all the patience from their owners, typically teenage girls, to keep them happy. Early versions of Cognitive Things remind me of the half a dozen Tamagotchis my daughter was maintaining as pets while she was in the middle school. Each night I am working hard to charge all my devices, which require batteries in different sizes and shapes. When I travel, I must pack all the different cables covering all of the devices, or face their non-working status when I need them the most on the

[1] Melissa Batchelor Warnke, "Why We Were Addicted to Our Tamagotchis", Vice, July 19, 2015, http://www.vice.com/read/in-praise-of-tamagotchi-683

A. Sathi, *Cognitive (Internet of) Things*,
DOI 10.1057/978-1-137-59466-2_8

road. Under Armour gave me the first ray of hope, as the running shoes are Bluetooth-enabled with a battery that lasts the "life of the shoe".

The second thorny issue everyone is facing is training. Cognitive Things are good at repeating what they have learned tirelessly, but how do I teach them to learn tirelessly from the changing environment, and how long would it take to train a cognitive thing before it can perform with some common sense. In most interactive systems, it is important for the cognitive thing to comprehend many variations of the sentences and comprehend their intent, especially as they begin dealing with tone, accent, and language differences. How are Cognitive Things bridging the learning gap and can they be faster learners of new subject areas?

Privacy and security represent the next area of concern. All the apps would like to know my current location and share it across those apps. Is it safe? Am I going to be stalked by a number of marketing things, which push campaigns at me at every available opportunity? Am I likely to be robbed by a fraudster? Are the Cognitive Things designed to give me options and what are the best options to use for my personality and risk level? If a cognitive thing offers to synch up with my Facebook account, should I let it do it and gain in integration or deny it to preserve my privacy?

How will Cognitive Things be organized? Will each of them have their own autonomy, free will, and ability to represent the organization they represent. Are we dealing with the shared pool of experts in a centralized organization, capable of centrally organizing and standardizing the entire organization? While on one hand the decentralized organization seems to represent the politics of the large corporation, on the other, centralized organization seems like an effort to create a big brother to watch the independent business units.

Cognitive Things will require a fair amount of space and many ups and downs in the utilization of the computing resources. There may be economies of scale for third parties to maintain content extraction, governance and application development. Should these Cognitive Things be placed in a public cloud, private cloud, or close to the data sources in the current data centers? Physical storage in a data center may provide security for customer data. Public cloud may be the best way to share the costs and maintenance across many organizations. While there is a trend to move everything to public cloud, will this expose organizational assets to third parties?

As Cognitive Things get introduced into the production environment, these are representative questions that differentiate successful mainstream implementations from demonstrations and pilots that wow the enthusiasts. Careful experiments with various options will give us better understanding of use cases where the decisions matter.

8.2 Installation and Maintenance

Last year, my wife and I were building a new house in Southern California and were presented with a myriad of security options. Unable to make a decision, we took the path of least resistance, laying the cables for a future security system, without making all the expensive equipment decisions. A year later, as my father needed monitoring, we asked the home security provider to install additional cameras and connect them to the Wi-Fi. Given that all the circuitry and wiring was already in place, we expected the task to be a simple one. Contrary to our prediction, two installers struggled with the installation for about eight hours and had to return the next day to finish the task. The good news is I now have a set of cameras watching the house; the bad news is it was not a simple task to install and configure.

Why is the installation of Cognitive Things so challenging? In some industries, manufacturers are still in the early generation of development, where the IoTs require a significant amount of configuration. Added to the trouble, these tasks are all new to the current work force and they have not been provided enough configuration and diagnostics tools. In order to get wide spread deployment, the Cognitive Things must become easy to configure. My Fitbit provided a comparatively simpler configuration option. It merely needed to be paired with an Internet connected device capable of connecting with the Fitbit via a Bluetooth. How is it that some Cognitive Things are a breeze to install, while others take a lifetime? If each of the 25 billion IoTs took two hours to configure, the human race will require 50 billion installation hours, or roughly seven for each human.

An easy installation involves a relatively small number of interface points and uses interface standards and wizards to mechanize and sometime automate as much of the interface configuration as possible. A closed ecosystem, guided by a single organization drives simplified interfaces. However, security systems and home automation are good examples of situations where it is hard to standardize interfaces.

On the flip side, once configured and in use, Cognitive Things can give us plenty of time saving through improved preventive maintenance. Ten years ago, my cars provided no heads up towards preventive maintenance. I was expected to get my car maintained every 7500 miles and a reminder from the car dealership showed up in my mailbox once every six months, based on their prediction that I was driving the car about 15,000 miles per year. My latest cars are somewhat better. They are able to predict maintenance based on their analysis of the usage and performance data from the car components and they also provide me with warnings for other things, such as tire pressure.

So, how would the dream cognitive thing maintain itself? There are three important cognitive activities:

1. *Self-Monitor*: Their ability to observe the behavior of their parts, analyze trends, and predict trouble spots. The cognitive thing must have the ability to differentiate the deviation from the norm and when to trigger action. On one hand, they should use simple fixes to avoid major failures, while on the other hand minimizing the effort required by the owner to repair unnecessary fixes.
2. *Self-diagnostics*: Their ability to use the performance data and a body of troubleshooting knowledge to isolate a malfunctioning component, replace, or reconfigure, if possible.
3. *Schedule and negotiate maintenance*: Their ability to apply usage data to decide when to repair to reduce outage. In performing this scheduling, the cognitive thing will negotiate with the schedule of the repair organization as well as the user to find an appropriate time suitable to everyone, without jeopardizing an outage.

Preventive maintenance is a growing expertise area. For decades, buyers received owners' manuals from the manufacturers. These manuals carried commonly known troubleshooting procedures. Are these manuals getting periodically updated as new types of problems are discovered and new service patches applied to the software components? Also, manuals carry static versions of troubleshooting procedures. Can there be an interactive version, which uses evidence and recent tests to prune the tree and offer the most likely investigation, test and repair process.

In the most sophisticated Cognitive Things, the events collected can give rise to prediction failures. The troubleshooting knowledge can be initially generated using prescriptive troubleshooting processes, often created by the engineering departments. As the associated equipment gets used, the troubleshooting process is used by all the users, and their successful paths can be crowdsourced to update the originally established prescriptive troubleshooting process.

8.3 Training

What is the knowledge quotient for your cognitive thing? In most consumer-facing situations, a cognitive thing can only be introduced if it has adequate knowledge out-of-box and does not work at the level of an oxymoron. For the cognitive thing to behave like a normal cognitive being, it must have enough knowledge to not ask the simple questions, and to comprehend most common situations—which may be relatively simple for a person with training, but often a gigantic leap for an untrained cognitive thing. In most situations, Cognitive Things are offering decent off-the-shelf functionality. How do they achieve that and how do they keep up with changes as trends change?

Let us look at the virtual agent as an example. In a typical call center operation, it takes many weeks to train a new call center representative. As they learn and get better, they often get promoted to do better jobs within the company. Training in call centers is typically a fairly organized activity and many customer-facing organizations built robust processes for training new employees, with the anticipation that there will be nearly 40–50 % turnover of their call center staff each year. The good news is that such operations have good training documentation, updated regularly by the training organization. As these types of operations move towards Cognitive Things, what is the best way to transfer this experience and organization to training and to update the Cognitive Things?

In classic human training, each person enters the program with a fair amount of common knowledge. Training adds a layer of specific terms, processes, and rules. However, in most cases, students retain a small percentage of the information given to them and need to use on-the-job learning to relearn and reinforce the rest. Even with good retention, there is a difference between learning a process in a classroom, from on-the-fly navigating it with an irate customer. Representatives learn over time both the technical aspects of the conversations as well as the softer skills, such as listening and responding.

To emulate cognitive activities, Cognitive Things are being designed with training processes similar to training schools for humans. Cognitive Things are provided with a series of processes and are asked to conduct targeted activities in a simulated environment. However, training Cognitive Things is a little different. They remember everything they are taught, but do not have the same ability to apply common sense or creative thinking while facing an explosive situation. Depending on the processes, task sophistication, and subject area, the training may need to include additional background material, ways of dealing with exceptions, and where to find additional material. On-the-job learning is different too. Cognitive Things can meticulously log each step, their success or failure across the entire population of interactions, and analyze it to improve interactions in future. However, they are not creative, so the training must explicitly deal with how to respond gracefully to unknown situations.

As I compare the training for Cognitive Things with training for call center representatives, a couple of observations can be made. First, the designer of a cognitive thing may consider pre-training a base level of common sense, such as knowledge about geography. So, if a USA-based customer were to make reference to a forthcoming trip to Paris, the cognitive agent must deduce this is regarding international travel. Second, the training must include exceptional situations and how to gracefully exit an infinite loop. You may have encountered stubborn Interactive Voice Response (IVR) units, which keep asking you for an account number, if you failed to provide one, and despite your desire to talk to a live agent, you get an endless cycle of "please punch your account ID one digit at a time". A cognitive thing can easily create a more cognitive example of the infinite loop, where the agent may sound very apologetic and nice, but be completely useless in understanding or responding to the intent of the questions. The third aspect is in using the conversation log to evaluate successful paths, improvements for failed paths, and new ways of dealing with a problem, discovered through the user interaction. Last but not least, training is an iterative process. It is a good idea to train the software, test its effectiveness in a simulated environment, evaluate its effectiveness against its success criteria, and repeat the iterations till the desired thresholds for success criteria are exceeded.

I used call centers as example, because that is an area where the parallels are more easily explainable. Other use cases may not be as clear cut. In situations where a cognitive thing is only augmenting human activities with additional information, the training requirement may still involve the listening skill but may not require dialog navigation.

8.4 Security and Privacy

Cognitive Things can observe the entire population of its scoped users tirelessly, collect the data from each individual, effectively discover many abstract patterns, and correlate those patterns to decipher social or psychological dimensions. While that is a boon for marketers, social researchers, engineers and operations, it is at the same time an exposure to fraud, a possible irritant for the consumer, and a regulatory risk item for societies and governments.

Misuse of data often results in consumer or regulator backlash. In many situations, manufacturers have had to curb their recording, learning and sharing of their operations. For example, PC World reported that LG Smart Televisions were recording viewing behavior, set as default, and continued

Cartoon 8.1 Dilbert's security
(Scott Adams, "Dilbert", February 28, 2016, http://dilbert.com/strip/2016-02-28, reprinted with permission)

to share viewing data, even after the option was turned off by the user.[2] Sprint Communications had to remove its device monitoring using Carrier IQ in 2011.[3] The use of phone company data by US National Security Administration and police to track terrorists and other criminals is a hotly discussed and debated item.[4] Sunil Agarwal and Ajit Ninan have vivid imagination and technical background. Cartoon 8.2 depicts the scenario portrayed by them as kitchen appliances get connected to the Internet (for the non-Indian audience, Amul is a famous Indian brand of butter).

Sometime extremes are not the best options. Privacy is one of those attributes where individual preferences vary, and a blanket decision may not fully represent the population. Would a teenager be willing to reveal her identity for a chance to appear on a television display in a concert? Some would be more than delighted, but a shy one will never repeat the experience and may not show up for the concert the next time. Can marketers capture the privacy preferences, test the ideas, and get confirmation to ascertain consumers' willingness, and then remember and apply the choices? If I receive a toaster for free, would I be willing to eat the toast with an Amul print on the bread? It

[2] Lucian Constantin, "LG smart TVs send data about users' files and viewing habits to the company", PC World, Nov 22, 2013, http://www.pcworld.com/article/2066400/lg-smart-tvs-share-data-about-users-files-and-viewing-habits-with-the-company.html

[3] Elnor Mills, "Sprint disabling Carrier IQ on Phones", CNET, Dec. 16, 2011, http://www.cnet.com/news/sprint-disabling-carrier-iq-on-phones/#!

[4] John Kelly, "Cellphone data spying: It's not just the NSA", USA Today, Dec. 8, 2013, http://www.usatoday.com/story/news/nation/2013/12/08/cellphone-data-spying-nsa-police/3902809/

Cartoon 8.2 Would you like to receive pop-up ads from your toaster?

tells us if alternative business models can be used for an increasing user base for a device.

Security is a different and bigger concern. Consumers are happy to trust their providers, until they receive a letter from the provider apologizing for an accidental leak of their credit card details to fraudsters worldwide. Unfortunately, Cognitive Things exacerbate the security exposure problem. With so much information collected about consumers and analyzed to get minute details about hangouts, usage and preferences, the data becomes a gold mine for the thieves and a liability for accidental leaks. As everyone has experienced, the security problem is shared across most major consumer brands. Are they ready for more cognitive insights to be stored and leaked from their data centers?

One solution to the security issue is to mask the data early and often. Data masking may retain the statistical value of the data, but may obfuscate the customer Personally Identifiable Information (PII). Further, query rules can be set in a way to avoid unique correlation across attributes to guess the identity of a single customer ("the adult males over 40 who are the residents of 1600 Pennsylvania Ave, has second hangout at Martha Vineyard and often travel to international locations"). In most situations, limiting the query to 25 or more individuals is a safe assumption, so if the response retrieves less than 25 individuals, the response is blocked.

8.5 Centralized or Distributed Architecture

As organizations acquire Cognitive Things, how will they be organized? Will they be centrally managed, distributed across the organization, providing a cognitive addition to each employee, or end up creating new organization structures? As Cognitive Things support inter-organizational communication, will they be neutral third parties, resellers tied to a single or multiple organizations, or an extension of the organization they represent? How would a distributed organization of Cognitive Things impact their information gathering, goal sharing, and protection of organizational and personal privacy? Will these Cognitive Things form a society of their own, sharing expertise across a diverse set of cognitive experts, even outside the normal span of collaboration for the human organizations they represent? How will they share knowledge, assign intellectual property rights, conduct auctions, or conduct voting to facilitate collaborative action? These questions may look far-fetched today, but they have been around for a while, leading to an entire branch of artificial intelligence aptly termed "Distributed AI".[5]

Distributed processing has its unique challenges. As described in Chapter 6, Cognitive Things create a lot of data. Analytics at the edge is the best way to resolve the data tsunami problem. However, analytics at the edge brings many new opportunities as well as related delineation of corporate goals and knowledge ownership to the forefront, many of them heavily debated in the distributed Artificial Intelligence literature. Source of data and source of analytics processing may not always belong to the same legal entity, thereby creating a set of valuable assets to be traded and monetized. Let me use an example to illustrate. Phone usage and location data belongs to consumers, with device manufacturers and telecom service providers as trusted keepers. The advertisements exposure may get stored in data management platforms (DMP) and may belong to a media service provider; the past purchase data may be collected by Nielsen or Slice. A consumer products marketer may have developed sophisticated analytics, which requires all this data, but pieces of the analytics engine may need to run in different edge environments—at the telecom service provider, DMP, or Nielsen. To safeguard consumer trust, the data owner may impose certain privacy rules on how the raw data can be accessed. The rules for advertising saturation and real-time bidding may be the property of the demand-side platform

[5] Les Gasser and Michael Huhns, "Distributed Artificial Intelligence", Volume I and II, Pitman London, 1989, https://www.amazon.com/Distributed-Artificial-Intelligence-Research-Intelligenc/dp/1558600922/

(DSP), which must be applied in placing the advertisements on the content the consumer is viewing. The segment selection and links to product promotions are generated at the consumer marketing organization. A set of distributed engines may use the data and these rules to select advertisements during media viewing. Many of these actions may take place closer to the edge, implying scoring rules must be sent to the edge application to execute the logic. Alternatively, carefully filtered data or related insight may be transmitted to other organizations, where they can use the data to make decisions. In either case, the logic is applied using cognitive agents who interpret the data using the processes given to them. Figure 8.1 shows this distributed architecture where the components are owned and operated by many diverse organizations, and come together for a collaborative decision-making in presenting a promotional offer to a consumer, consistent with their location, usage, and privacy preferences.

I chose not to draw the many interconnection points assumed in this picture. However, in a near real-time environment, collaboration across these systems requires excellent interfaces across many organizations, and ability to share data and logic.

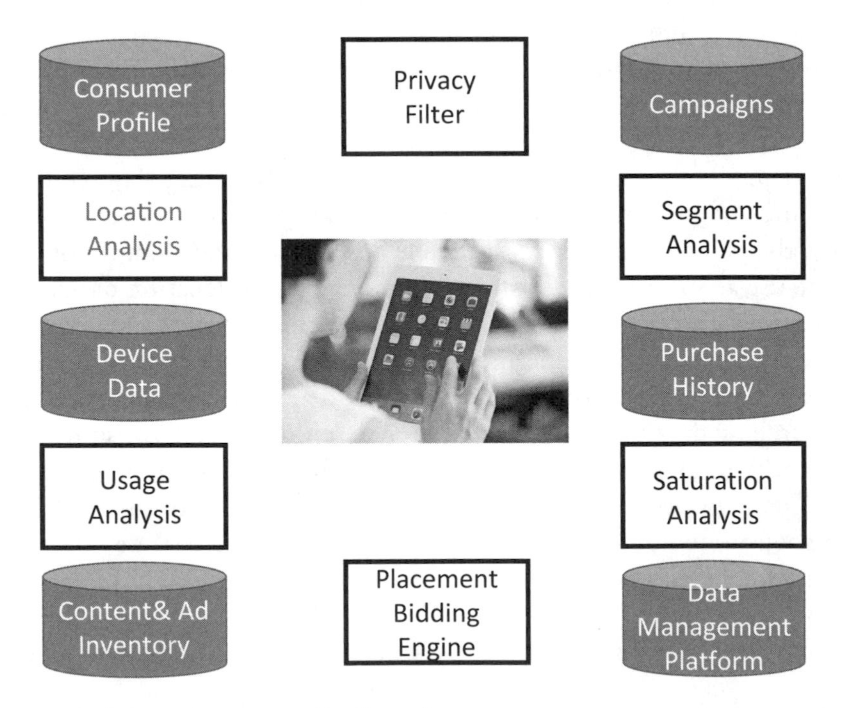

Fig. 8.1 Collaboration across distributed sources

8.6 Cloud or On-Premise Infrastructure

Cognitive Things require many services, all the way from structured to unstructured data, from data acquisition and prep to decision-making and execution, and from a variety of sources that come together in near real-time. Cloud technology has offered an attractive proposition to anyone who is seeking to build Cognitive Things using a number of atomic services, which can be integrated for specific use cases. Bluemix is IBM's flavor of Cloud Foundry, an open-source software development platform. The techie term for that category is "platform as a service". Developers at companies can build, test, and deploy custom applications using this set of software and services. One selling point of Cloud Foundry is that it purports to work across cloud infrastructures. The PivotalCF version, for example, runs on Amazon Web Services and VMware's vCloud Air.[6]

Time-shared computing was popularized by mainframe computers, which were expensive to buy and house. Getting leased access to the mainframe was the best way of using it without incurring massive capital and other fixed costs. Clouds apply the same principle to the modern computing infrastructure. A variety of cloud solutions have emerged over the last decade. Public clouds house data across many corporations or consumers and offer secure access to each. Private clouds house the data within a corporation's firewalls, but use the cloud infrastructure to reconfigure the environment for each project, thereby reducing the dedicated purchase of computing infrastructure for each project. Cloud technology can be used at different levels of computing infrastructure. An infrastructure cloud offers a computing environment. A storage cloud offers capabilities for storing data as well as capabilities for backup and restore. For example, Symantec and Amazon provide storage clouds for personal computer data backup. An application cloud houses an entire application, such as SalesForce.com. One big issue for businesses in the cloud computing era is making sure that certain data and applications run in certain places for regulatory compliance or other reasons. IBM's Bluemix Local offering will enable that by letting business customers write their business applications once and then run them across a full complement of clouds worldwide and/or in their own data centers.

---

[6] Barb Darrow, "IBM puts Bluemix everywhere", Fortune, October 1, 2015, http://fortune.com/2015/10/01/ibm-bluemix-local/

As large quantities of IoT data started to emerge, cloud providers offered ready-to-use solutions for analysis of this data. It is an easy decision to use a public storage for already public data and analyze it for specific queries. Cloud providers offered low entry points and subscription fees to simplify starting costs for analytics. Weather services from the The Weather is a great example of this data. The combination of the IoT and cloud computing will enable more than 100,000 weather sensors and aircraft along with "millions of smartphones, buildings and even moving vehicles" to combine information. The weather service is based on data from thousands of sources, resulting in approximately 2.2 billion unique forecast points worldwide. It averages more than 10 billion forecasts a day on active weather days.[7]

The pricing models for cloud-based cognitive services have significantly challenged the software and services industry. In a typical capital purchase, software is sold with an upfront fee and an annual maintenance fee, and customization services are also front-loaded and may cost as much as, or more than, the software. For a typical traditional automation project, nearly 50 percent of the cost may need to be incurred in the first year, while most of the benefits may be back-loaded. To make these programs viable, most of the large programs were capitalized with a multi-year amortization schedule. The cloud changed the model to a monthly subscription, where the costs are mostly transaction-driven and, hence, back-loaded, while the benefits may be accelerated through early deployments. In addition, the funding organization may choose to cancel the program any time without incurring expensive upfront costs.

8.7 Chapter Summary

In this chapter we have looked at many implementation issues: installation, maintenance, training, security and privacy, distributed processing systems, and hybrid cloud-based computing environments. While many of these issues are shared with any automation project, there are a couple of unique aspects associated with Cognitive Things:

- The sheer magnitude and variety of consumer-facing Cognitive Things make installation an important show-stopper for success. Consumer-facing

[7] Collin Barker, The cloud for clouds: IBM and The Weather Company work on big data weather forecasts, ZDNet, March 31, 2015, http://www.zdnet.com/article/the-cloud-for-clouds-ibm-and-the-weather-company-work-on-big-data-weather-forecasts/

Cognitive Things must be designed for easy installation and maintenance. If it takes hours to install a cognitive thing and is a constant drain on resources and takes time to maintain, the adaption beyond innovators and early adapters will be a tough sell unless there is a compelling use case. The smartphones, tablets, and watches are interesting exceptions as high maintenance Cognitive Things, which have already moved well into late adapters.
- Training for Cognitive Things includes subject matter knowledge, which must be acquired and changed with learning and environment changes. Cognitive Things offer great capabilities for observing and logging usage, which can be used for speedier learning.
- Security and privacy is a bigger concern because of the extent of insight that can be shared and inappropriately used. Privacy is a personal choice, and consumers would like to be in charge, where they can barter privacy for discounted or free services. However, security is a bigger concern and can easily overshadow willingness to share the data.
- Cognitive Things are inherently distributed, and require many aspects of distributed computing. In addition to the distributed processing requirements, data and knowledge ownership and partial sharing based on use case is a new add-on.
- Hybrid cloud is emerging as the favorite for computing infrastructure, as it offers cost-effective sharing of services, while protecting PII information.

The last three chapters (6, 7, and 8) provided a view of the technical implementations. The next four chapters will focus on how Cognitive Things interact with the world and impact the environment around them.

9

Machine-to-Machine Interfaces

9.1 Introduction

Every time I am in the streets of Delhi, I am subjected to noise pollution created by a bunch of aggressive drivers, and often wonder if that would be a relic of the past in a couple of decades? Imagine a highway full of self-driven cars. As a self-driven car yields to another car, how are they likely to communicate with each other? How often will the self-driven cars honk or blink their headlights? In the case of a traffic jam, how would they decide who gets priority and who yields? Also, if they are driving for a long time, will they maintain a safe distance from each other, or hitch with each other like cars in a train?

As Cognitive Things multiply in number, they will begin to deal with each other more often than dealing with humans. How will their communication be organized and how will it differ from human conversation? In this chapter, we examine the basic components of machine-to-machine communication. It would be good to identify these components before we deal with more difficult conversations, as they start involving humans.

Machines have some important characteristics that differ from humans. They can repeat a fair amount of conversation over and over without getting tired. In fact, repetition makes it easier for them to maintain context in interpersonal communication. Machines can be a lot more explicit in their communication. Sharing across many machines is a lot easier. A conversation can easily involve a large number of machines, as long as there are well-defined rules of conversation. Machine-to-machine communication has been a topic

A. Sathi, *Cognitive (Internet of) Things*,
DOI 10.1057/978-1-137-59466-2_9

of intense research among Distributed AI researchers.[1] Now the technology is finally catching up with the research, giving an opportunity to implement machine-to-machine collaboration.

We will look at four aspects of machine-to-machine communication. First, how do machines connect with each other and which language do they use for communicating with each other? A number of competing technologies have emerged for transmitting information from one cognitive thing to another. This section summarizes the alternatives and compares them using engineering concerns such as distance between devices and energy requirements. Second, how do they identify each other and establish relationships? Most of the conversation uses wireless technologies, often in open areas. Machines need to understand who they are communicating with and establish a connection so that they can communicate without eavesdropping by others. Third, I will discuss issues related to information governance, such as data quality, master data management, and information life cycle management. Many of these devices operate at the edge using limited storage and battery. Sending large amounts of information may not be the best solution. Often, machines carry and process local information and share small amounts of insights. How are these machines governed to optimize local and shared data? Fourth, how do they negotiate with each other? Machines are already performing tasks on behalf of their human or organizational masters. They must defend goals, follow the policies laid out to them, discuss alternatives with other machines, and come to a favorable conclusion, hopefully resulting in a win-win. Machine-to-machine negotiation is a hard cognitive skill, which must be mastered by the machines if they are to truly collaborate in the real world.

9.2 Communication Media

Ever since humans have evolved their cognitive skills, communication with others became an important aspect of cognitive behavior. Humans learned many ways of communicating, including paper and pen, smoke and fire, sound and music, wires and electrons, and many more. It was natural that as the Internet of Things evolved, the Cognitive Things would find more than one way of communicating with each other. This section provides a short summary of these communication mechanisms and discusses their pros and cons.

---

[1] A. Bond and L. Gasser. Readings in Distributed Artificial Intelligence. Morgan Kaufman, San Mateo, CA, 1988, https://www.amazon.com/Readings-Distributed-Artificial-Intelligence-Alan/dp/093461363X

As communication engineers describe their work, they often refer to the Open System Interconnect (OSI), which is a joint standard that resulted from independent projects by the International Organization of Standardization (ISO) and the International Telegraph and Telephone Consultative Committee, or CCITT. It was published in 1984 by both the ISO, as standard ISO 7498, and the renamed CCITT (now called the Telecommunications Standardization Sector of the International Telecommunication Union or ITU-T) as standard X.200. In this model, a networking system was divided into layers. Each entity interacted directly only with the layer immediately beneath it and provided facilities for use by the layer above it. Protocols enable an entity in one host to interact with a corresponding entity at the same layer in another host. The seven layers are described in Table 9.1.[2] The first layer represents the physical infrastructure including the fiber for wireline networks or the radio waves for the wireless networks. The second layer represents the reliable transmission of data on the physical network. The third layer represents the network, including routers and switchers that connect the physical network and establish routing of packets. The fourth layer represents transport of data using the network established by the third layer. The fifth layer manages a session, which may involve many transports between the connected nodes. The sixth layer provides ways to transform, encrypt and compress the data for its secured and efficient transmission. Lastly, the seventh layer deals with applications and facilitates the use of the layers by applications like email and web browsing.

Each of the methods of communication for Cognitive Things crisscross these seven layers, most often covering some but not all of the seven layers, and hence requiring additional mechanisms for end-to-end communication management. Let me go through a couple of them. For further details on this topic, I would like to refer the reader to an excellent book written by Michael Miller.[3]

TCP/IP The de facto standard network protocol today, used for both Internet and local area network (LAN) connections is called TCP/IP (Transmission Control Protocol/Internet Protocol). The IP part of this protocol provides the standard set of rules and specification that enables the routing of data packets from one network to another (network layer in the OSI stack). After the packet reaches the receiving end, TCP translates the packet into its original

[2] Jim Geier, "Wireless System Architecture: How Wireless Works", chapter from the book "Wireless Networks First Step", Cisco Press, August 2004, http://www.ciscopress.com/store/wireless-networks-first-step-9781587201110

[3] Michael Miller, "The Internet of Things: How Smart TVs, Smart Cars, Smart Homes, and Smart Cities are Changing the World", Que. Publishing, March 2015, ISBN: 0789754002, https://www.amazon.com/Internet-Things-Smart-Cities-Changing/dp/0789754002

Table 9.1 OSI layers

Layer	Description
Application	It establishes communications among users and provides basic communications services such as file transfer and e-mail. Examples of software that runs at this layer include Simple Mail Transfer Protocol (SMTP), HyperText Transfer Protocol (HTTP) and File Transfer Protocol (FTP).
Presentation	The presentation layer translates data between a networking service and an application; including character encoding, data compression and encryption/decryption. For example, this layer can translate the coding that represents the data when communicating with a remote system made by a different vendor.
Session	It establishes, manages, and terminates sessions between applications. Wireless middleware and access controllers provide this form of connectivity over wireless networks. If the wireless network encounters interference, the session layer functions will suspend communications until the interference goes away.
Transport	Provides mechanisms for the establishment, maintenance, and orderly termination of virtual circuits, while shielding the higher layers from the network implementation details. In general, these circuits are connections made between network applications from one end of the communications circuit to another (such as between the web browser on a laptop to a web page on a server). Protocols such as Transmission Control Protocol (TCP) operate at this layer.
Network	It establishes path determination and logical addressing. The network layer provides the routing of packets though a network from source to destination. This routing ensures that data packets are sent in a direction that leads to a particular destination. Protocols such as Internet Protocol (IP) operate at this layer.
Data Link	The Data link ensures reliable transmission of data frames between two nodes connected by a physical layer. It manages synchronization and error control between two entities. With wireless networks, this often involves coordination of access to the common air medium and recovery from errors that might occur in the data as it propagates from source to destination.
Physical	The Physical layer provides the actual transmission of information through the medium. Physical layers include coaxial cable, optical fiber, radio waves and infrared light.

format and combines multiple packets back into a single file (transport layer in the OSI stack). For the TCP/IP to work, each device is assigned an IP address (see Section 9.3 Identity Management for more details on IP naming).

Wi-Fi Wi-Fi (Wireless Fidelity) networks are the most popular wireless networks for homes. They utilize a hub and spoke configuration, where all the devices are connected to the router, and communicate with other devices using the router as the hub. Each of these devices has to be manually configured to connect to the network. In order to keep my house secure, I must set a password for my Wi-Fi. If each Cognitive Thing in my house were to be connected using Wi-Fi, I will need to provide my Wi-Fi password to each device. This is certainly not feasible if the number of devices were in the hundreds; for example, in the case of an intelligent lighting system in a house. One option is to develop a hierarchy, where small sensor devices are connected to a master device that can then connect to the Internet via Wi-Fi.

Bluetooth Often IoTs are connected using Bluetooth which uses peer-to-peer network connectivity. Unlike the hub and spoke model of Wi-Fi, Bluetooth facilitates a direct connection between two devices. Bluetooth radios are extremely small, making them ideal for miniature IoT sensor devices. Cars and smartphones extensively use this technology for connection and auto-synchronization. To set up a Bluetooth connection, both peers need to accept the connection.

Cellular Networks Wireless service providers offer public networks using cellular technology. The cellular networks establish adjacent cell sites with one or more fixed-location transceivers that provide coverage to a geographical area. Each cell uses a different frequency than its neighboring cells so that the signals do not interfere. Cellular networks offer a number of desirable features:

- More capacity than a single large transmitter since the same frequency can be used for multiple links as long as they are in different cells.
- Mobile devices use less power than with a single transmitter or satellite since the cell towers are closer.
- Larger coverage area than a single terrestrial transmitter, since additional cell towers can be added indefinitely and are not limited by the horizon.

9.3 Identity Management

As with humans, machines are sometimes competing but also sometimes collaborating with each other. As they find an opportunity to work with each other, they must verify each other's identity as well as their roles and relationships before engaging in communication. It allows them to establish what they can share, whether they are allowed to be transacting with each other, and whether they should be collaborative. In some situations, machines are already set up in a hierarchical organization. For examples, the sensors and gateways connected with a centralized cloud server are organized in a hierarchy, and once the trusted communication is established, they can request and respond without any constraints based on the software installed. In other situations, such as iPhones with various IoTs connecting them via Bluetooth, the authentication allows them to share Internet accessibility and remote commands. In the case of Wi-Fi hubs, administrators and users may have different privileges. Finally, cars on the road communicate their intention to change lanes, apply the brake, and so on, and may receive permission from other self-driving cars yielding to them.

In all of these situations, the Cognitive Things are expected to communicate with other things, authenticate each other before communicating, and establish roles and rights. Locks and keys were an important invention in organized society to protect assets. As computers were introduced in the automation era, they borrowed the concept of lock and key and transformed to user IDs and passwords. I have over a hundred passwords in my keychain, and I have trouble keeping track of whether I have a password for a specific site, let alone the details, which depending on the site rules, may require capitalized letters, numbers, or special characters.

However, locks and keys and the corresponding user IDs and passwords are not very cognitive. They are very impersonal and are very open to the person carrying the key or the password. Depending on the implementation, they may have a set of rules and people to support those miserable people who frequently forget their passwords. As machines start to communicate with each other, should they create hundreds of passwords and store them in their vaults? They will require key chains for password management, process for reseting passwords, and be subject to security risks. Alternatively, machines could use other ways of recognizing other machines and identify rules of engagement.

Let me start by discussing what is the objective of the identification and authentication process and then introduce how cognitive computing can reduce the pain of managing passwords. In a typical machine-to-machine communication, they will be expected to deal with three types of relationships:

1. Trusted Cognitive Things: These Cognitive Things have been established as a trusted peer, maybe because they belong to the same organization. The cognitive thing must share openly and freely. This could include all the home appliances as they deal with the home security system, home management system, or the cognitive assistant who is working on maintenance visit scheduling.
2. Strangers and acquaintances: These Cognitive Things are not yet trusted, but may communicate based on established societal laws and regulations. The communication should only include publicly sharable information.

A self-driven car may publicly share a left turn signal signifying its intention to turn left at the next intersection with other cars on the street.
3. Bad Cognitive Things: Like villains in the human world, these Cognitive Things are suspected to be engaged in prohibited activities and should be avoided at all costs. If a network element finds that a line provisioned for vending machines is downloading popular movies, this may involve fraudulent data communication activity.

IoT actors authenticate by presenting security tokens on their calls or messages to each other. Tokens represent the relationship between the relevant user and the calling actor (and any consents or permissions associated with that relationship). Upon receiving a message, an actor validates the token to verify that the request is consistent with the relationship or permissions. If consent is removed, the token is revoked, and access is disabled. OAuth 2.0 & OpenID Connect 1.0 are two authentication and authorization frameworks that enable this model.[4] An LDAP (Lightweight Directory Access Protocol) server may be used to keep track of all the valid things, classified into the three categories above.

The original IP addresses used IPv4 which offered approximately 4.3 billion unique addresses. In preparation for the 25 billion IoTs, the new format, IPv6 uses a 128-bit address, theoretically allowing 2^{128}, or approximately 3.4×10^{38} addresses.

Much like finger printing, voice or facial recognition for humans, machines can identify each other with their IDs and past behaviors. Trusted relationships can be established with friends, and firewalls can be used to establish rules for sharing. Public information can be shared with strangers. Machines can remember other machines through their IDs and may treat them as acquaintances until an event causes them to reclassify them as either trusted friends or dishonest fraudsters.

9.4 Information Governance

In this section, we will look at three major information governance topics: data quality, master data management, and information life cycle management. Cognitive Things deal with large amounts of data. For their

[4] John Bradley, "An Authentication Framework for the IoT", Slideshare, March 10, 2015, http://www.slideshare.net/AllSeenAlliance/identity-foriot03

proper performance, it is important to establish good governance practices. Failure to do so could result in major malfunctions, or breakdowns.

Like humans, machines may sometimes lose the communication channels, fail to communicate, or otherwise lack information needed for communication. Loss of data and associated data quality assessment is an integral part of communication. If a communication is lost, machines try to re-communicate or may in some cases deal with less than perfect information. A turn signal signifies intent to turn. A turning vehicle without a turn signal poses a traffic challenge, which may be originated by a human, a malfunctioning turn signal, or lack of visibility. In either case, others vehicles must adjust to less than perfect data communication in a similar way to how humans drive cars currently.

So far the assumption was that data quality issues arise from missing information. How about dealing with misleading information or conflicting priorities? Just because a cognitive thing is supplied by a manufacturer and is programmed to deal with communication with the manufacturer, it is not always the case that the manufacturer has the best interest of the service provider or the customer. For example, a smartphone may represent many priorities resulting from an operating system supplier, the smartphone manufacturer, the service provider, the corporation paying for the phone and the employee using the phone. If there are conflicting sources of instructions, how does the smartphone decide whom to listen to? If the phone is stolen and then reused, the new user would certainly not have the best interest in mind. In a situation where a specific service provider has changed the service process for their best customers, manufacturer-created default processes may not be the best way to proceed.

The ownership organization needs to be explicitly understood by the server supporting the network of Cognitive Things. In the human organization, master data management solutions maintain a customer hierarchy, which govern the rights and responsibilities of humans dealing with a shared set of assets. It establishes the price, the product ownership, and the authority of individuals who can initiate or receive certain actions. Changes to ownership data can be governed using a distributed governance process and documented using blockchains. As Cognitive Things step into human roles, a master data of things maintains their organization, ownership structure, authority, and responsibility. In a typical home automation solution, the system manages the master data of all the Cognitive Things operating in the house, their service providers, manufacturers, and rules of engagement, specifying who has access to the cognitive thing. If the battery of a driverless car is supported by a third party, the car management system must have information on how to find the battery service provider and how to deal with a non-functioning battery when the car travels outside the designated service area for the battery service provider.

Cognitive Things generate a lot of data and deduce a fair amount of insight. They are certainly more capable of storing and retaining a lot of data, but how much should they retain, and how should they decide on what to forget? I may have liked my coffee with sugar when I was younger, but I may begin restricting sugar as I age. Human preferences change over time. How would Cognitive Things keep up with these changes and keep them consistent? Information life cycle plays a major role in keeping the data and insight recycled, so that the machines keep up with the changes.

Information life cycle management is also an important aspect of information governance in purging old data. As new data is received, information retention policies dictate how much data needs to be retained.

9.5 Negotiation

As I was working on my Ph.D. thesis in the early 1980s, my emphasis initially was to automate an expert using a knowledge-based reasoning system. However, ARPANET was just being introduced to connect computers, and researchers were very excited to share information with others connected via ARPANET giving me an opportunity to explore distributed communication opportunities. As I expanded my Ph.D. thesis to include distributed problem-solving, I found myself writing processes and algorithms for negotiation among intelligent agents. I found a ton of academic research on multi-agent negotiation. However, most of the modeling was about human negotiation. With the primitive state of inter-computer communication, the idea of connecting computers to negotiate with each other seemed far-fetched. I distinctly remember a major brainstorm session with Mark Fox, where he projected a vision of the world in which machines will have their own goals and will negotiate with others. It took the world about 30 more years to catch up with him, but it did give me an early peek at cooperative negotiation and distributed artificial intelligence.

What is negotiation in a machine-to-machine world? Reid Smith, a scientist working for Schlumberger, designed Contract Net, which provides a communication and negotiation mechanism among machines.[5] This research

[5] R.G. Smith, "The Contract Net Protocol: High-Level Communication and Control in a Distributed Problem Solver", IEEE Transaction on Computers, Issue No.12—December (1980 vol. 29), http://www.computer.org/csdl/trans/tc/1980/12/01675516-abs.html

work gave me a nice starting point for designing a collaborative negotiation process among Cognitive Things. In a classic multi-agent environment, there are multiple intelligent entities, each with their goals, constraints, and resources. They can make contracts with each other and need to decide what makes sense for them individually to agree to transact with each other. In a typical buyer–seller interaction, the buyer would like to purchase a good or service, while the seller has that good or service, and would like to obtain the best price for it. If this is a one-time purchase like residential real estate, the buyer would like to bargain for the cheapest price and the seller would like to maximize the price. The seller must keep the buyer engaged or risk losing the buyer and incur the cost of finding the next buyer, while the buyer would be concerned about other buyers in the market place. In a cooperative negotiation problem, agents must deal with each other over a number of contracts, such as spouses in a married life. In a typical cooperative negotiation, agents are more likely to share goals and constraints and find win-win solutions.

Cognitive Things require many types of negotiations. Two driverless cars sharing a busy road must decide right of way. Unlike two crazy rash drivers trying to cut out the last second from their travel time, the cognitive cars may follow the local street rules for their negotiation, and in some cases, for example in a traffic jam, may need to deviate from the rules for collaborative gains. A cognitive dishwasher seeking preventive maintenance may juggle through hard and soft constraints without seeking the best time at the expense of the maintenance technician. In each of these examples, there is more than one agent, each with some information to share, and they must mutually decide on an action based on a negotiation. The action may include transfer of assets, sharing of common or public assets, or simply a commitment for a joint activity in the future. Humans are fairly good at negotiation, and it is a very cognitive activity that a child understands in the early years of mental development and progressively gets better at. How do machines negotiate with each other, and do they mimic humans?

Each cognitive thing has a goal. For driverless car, the goal is to transport the car from point A to point B in the shortest amount of time. In addition, they have many hard or soft constraints. Hard constraints cannot be violated, while the soft ones are preferences that would be nice to have. In a simple two-agent negotiation, the agents exchange or trade resources with each other. In a more complex multi-agent negotiation, there could be multi-way exchanges, where the trades cascade from one pair of agents to the next in a closed loop. One cognitive agent could get a specific time for an appointment if another agent were to give up a previous appointment, and find a new appointment that is better than the previous appointment. Human administrative agents are masters at these three or four way exchanges. If machines were to take over that task, they must learn how to work like humans. As I worked on my Ph.D. thesis, I was proud of my 1 MIPS Vax machine from Digital Equipment Corporation and had all the faith in the machines' ability to search faster and better. Alas, the human expert I was dealing with was an executive administrative assistant with decades of experience and easily performed far better than my machine in the first blind test. By the third iteration, I was beginning to get some traces of her wisdom, and her ability to focus on constraint space. She skillfully tackled the choices associated with the most constrained situations and had an organizational understanding of how to focus on the tightest constraints. My first generation negotiation algorithm merrily looked at all options and even after an entire night of search, could not come anywhere near her split-second results.

This experience prepared me for dealing with many real-world cooperative negotiation situations over the years. In a typical traffic negotiation everyone must use common road communication practices to gain acceptance from others. Drivers often use hand signals to allow others to pass, while the receiver of the favor responds by gesturing a thanks to the first driver. It would be interesting to see how car manufacturers establish negotiation communication protocol across driverless cars. A good negotiator does not only look at their own goals and constraints, but anticipates considerations for others, to maximize collaborative goals, such as traffic speed. Cognitive Things will hopefully do better than the Delhi drivers honking their way to their destination. While machines have worked their way to beat humans in chess and jeopardy, negotiation still remains a primarily human task, and machines have a long way to go.

9.6 Chapter Summary

This chapter has focused on four distinct aspects of machine-to-machine communication. These four aspects collectively represent the many capabilities needed for collaboration across machines. Without proper communication, Cognitive Things become islands of automation.

First, I discussed how machines connect with each other and which language they might use for communicating with each other. A number of competing technologies have emerged for transmitting information from one cognitive thing to another. This section summarized the alternatives and contrasts them with engineering concerns such as distance between devices, and energy requirements. Smartphones are rapidly emerging as the universal hub for Cognitive Things, and offer many conveniences not found in competing solutions.

The second topic centered around building proper acquaintances and trusted relationships. Most of the conversation uses wireless technologies, often in open areas. Machines need to understand who they are communicating with and establish a connection so that they can communicate without eavesdropping by others. I described different types of relationships, including civil behavior among strangers, where each Cognitive Thing assumes others are authorized to use common resources, such as public roads, as well as a responsibility to protect itself from malfunctioning or malevolent Cognitive Things. Trusted relationships are built using explicit exchange of identifications and authorization using policies.

Third, how do they govern the information stored and shared by them? I discussed three aspects of information governance—data quality, master data management, and information life cycle management. Machine-to-machine communication may suffer from data quality issues and Cognitive Things will provide resilient mechanisms to deal with less than perfect data quality. Master data management must include device ownership and organization. As Cognitive Things change ownership and get leased, blockchain may provide a mechanism for recording and tracking ownership changes.

Fourth, how do they negotiate with each other? Machines are already performing tasks on behalf of their human or organizational masters. They must defend the goals, follow the policies laid out to them, discuss alternatives with other machines, and come to a favorable conclusion, hopefully resulting in a win-win. This is a hard cognitive skill and one that must be mastered by the machines, if they are to truly collaborate in the real world.

In the next chapter, we will explore man-to-machine communication and in Chapter 11, we look at how machines participate in human-to-human communication.

10

Man-to-Machine Interfaces

10.1 Introduction

Last year, we welcomed the arrival of our grandkids, who will be part of the first generation to grow up with Cognitive Things. According to predictions from *BusinessWeek* and others,[1] my grandkids will be as interested in driving cars as we love horseback riding—a leisure activity as opposed to a day-to-day necessity. My wife and I have also started to explore children's toys with our grandkids. Most of the talking toys were interactive, but fairly boring. As we pressed the buttons, there were only limited options. How much time would it take for a kid to exhaust all options and look for the next toy? We were looking for a toy that would cognitively grow with our grandkids.

CogniToys™ from Elemental Path attracted our attention. CogniToys provides just the right touch of cognition in its interactions with the kids. The toy plays games with the child to touch on educational topics like spelling, vocabulary, math, and geography. More interestingly, it takes what it learns from conversations and from what parents tell it about their children during the initial setup, and incorporates that information into the games. The first CogniToys is Dino, a green or pink dinosaur. It features a large blue button on its stomach. When kids press it and ask a question, the toy connects to an online knowledge database to figure out the best age-appropriate answer to the question. The toy both understands the question and responds in human-like

[1] Keith Naughton, "Can Detroit Beat Google to the Self-Driving Car? Inside GM's fight to get to the future first", Oct. 29, 2015, Bloomberg BusinessWeek, http://www.bloomberg.com/features/2015-gm-super-cruise-driverless-car/

A. Sathi, *Cognitive (Internet of) Things*,
DOI 10.1057/978-1-137-59466-2_10

sentences, thanks to IBM's artificial intelligence technology, Watson. "You can ask it a plain-English question and get a plain-English answer," Elemental Path co-founder JP Benini said in an interview before the project launched. "If kids ask questions about soccer, it will have a counting game and count soccer balls," Benini said. "You could be in the middle of asking questions about soccer, but it might ask you, 'Can you spell field goal?'" Certain topics like religion and, of course, "Where do babies come from?" are called "Ask mommy" questions, because the toy is programmed to gently deflect them away (Fig. 10.1).[2]

The cognitive human interface is a growing field with many breakthroughs. Unlike computers from the automation era, the new expectation from a computer is to be flexible, interactive, and empathetic. In this chapter, we will explore a couple of topics related to human–computer interaction. First, how do cognitive things authenticate humans? While user IDs and passwords were easy to conceptualize and program because of their similarity to locks and keys, they also shared their biggest weakness—they could be broken, stolen, or guessed. New biometric measures are far more personal and harder to break. Second, how do cognitive things conduct their interactions with humans? As much as human conversations are beginning to bring data, music, video, pictures, and other mechanisms for expression, what are cognitive things doing to make their interactions useful? Third, I would like to discuss the role of emotions and tones. Cognitive things are increasingly becoming sophisticated

Fig. 10.1 CogniToys from Elemental Path

[2] Eric Johnson, "Cognitoys for Kids Uses IBM's Watson to Talk Like a Buddy", re/code, February 17, 2015, http://recode.net/2015/02/17/cognitoys-for-kids-uses-ibms-watson-to-talk-like-a-buddy/

in sensing human emotion. How do they respond to the emotion, and can that emotion be embedded in other ways? Fourth, how do cognitive things achieve their goals and get the best win-win solution designed in their negotiation with humans? Can they learn something from human negotiation and incorporate this into the conversations with humans to make the interaction goal- and constraint-directed?

10.2 Authentication

In the last chapter, we looked at authentication for machines and how a simple user ID and password may not be sufficient for scaling the authentication. Identification of humans by machines runs into even bigger problems, as humans are more prone to forgetting their passwords. The problems are exacerbated and lead to frustrations if they are asked to change their password frequently. A number of other means of authentication are gradually becoming popular (Cartoon 10.1).

Finger and palm prints have been a popular form of authentication and are considered reliable even for criminal prosecution. Australia currently houses the largest repository of palm prints in the world. The new Australian National Automated Fingerprint Identification System (NAFIS) includes

Cartoon 10.1 Remember the Password

print sets from 3.9 million people and has been used in the identification of 64,000 finger and palm prints on crime scenes.[3] The ANSI/NIST international standard for fingerprint data exchange makes it easy for police services to exchange fingerprint records across police forces when necessary.[4] During contract execution, collecting signer thumbprints has long been considered a way to not only identify signers and safeguard against forgeries, but also to protect the Notary from allegations of wrongdoing. It was for this reason that the state of California passed a law in 1996 requiring notaries to obtain journal thumbprints for signers of certain documents and why many notaries nationwide continue to record signer thumbprints in their official notarial journal.[5] Consumer technology has finally caught up with this form of authentication and smartphones have started to use fingerprints as a reliable form of authentication. In the case of Apple iPhones, finger print authentication can be used as a single sign-on across a number of apps running on the iPhone including payment authorization. Fingerprints do run into problems, with even the tiniest amount of water, or because of anything else stuck on the finger. Most authentication mechanisms offer password-based authentication as a backup and often require password authentication after certain instances, such as when the iPhone restarts.

Computer-based voice recognition is yet another authentication mechanism, which is very useful when a customer calls the service provider. By comparing the voice impression from the call with stored voiceprints, the call center software can authenticate the customer. There are two approaches to authentication. Either the software will ask the customer to read a specific sentence, or the software will authenticate using voice impression from the conversation. The technology is maturing in the latter approach, and increasingly it is possible to use conversational voice for authentication. Unlike earlier voice-based authentication, this new generation of technology is passive and transparent. The required voice verification data is acquired during the opening conversation between the caller and service center personnel, eliminating the need for any cumbersome authentication questions. The authentication process is text-independent and freeform. The customer does not have to state any specific words or phrases, a facet that may have previously led to

---

[3] Crimtrac, "Fingerprints", https://www.crimtrac.gov.au/fingerprints

[4] Brad Wing, "American National Standard for Information Systems—Data Format for the Interchange of Fingerprint, Facial & Other Biometric Information", NIST Special Publication 500-290 Rev. 1, December 2013, http://biometrics.nist.gov/cs_links/standard/ansi_2012/Update-Final_Approved_Version.pdf

[5] Kelle Clarke, "Notary Trends: Privacy Issues And Collecting Signer Thumbprints", Notary Bulletin Best Practices, July 23, 2015, https://www.nationalnotary.org/notary-bulletin/blog/2015/07/privacy-issues-collecting-signer-thumbprints

concerns that a voice recording of an individual's discrete speech could be used to fraudulently access an account. It uses a content, language, and accent-independent authentication process. The capturing of the client's voiceprint is unrelated to the content of the conversation; rather, the closeness of a match to the stored print is determined by subtly unique characteristics such as vocal tract length and shape, pitch, and speaking rate. The system initially records one or two conversations, extracts the voice features that are distinctive to the client, automatically creates a reference voiceprint and stores it in a secure directory. By its very nature, the speaker's biometric data is unique, making the technique more secure than any password-based or "challenge question" process.[6] Banks and financial services companies are already rolling out pilot programs to see how customers use and react to them. MasterCard is working on voice recognition. USAA is using a combination of facial and voice recognition, depending on circumstances, in which the member reads a phrase aloud to gain access to his or her account. In Australia, the federal tax office is using voice recognition for account access.[7] Voice authentication works well, as long as the speaker is able to speak normally. A heavy nasal or throat congestion can make voice recognition useless.

Iris scanning provides a much better identification as compared to voice, finger and palm prints. It is a highly data rich biometric measurement, which is well protected from the environment, highly unique for each individual, and stable over time. The table below provides a comparison of various authentication processes in terms of their relative accuracy.[8] Iris recognition technology is actually more advanced than systems typically depicted in movies. In films, we often see actors lean close to the scanners, practically placing their eyes on the screen. Then they stand perfectly still and wait patiently for identification. In reality, current iris recognition technology can identify users in less than a second from up to 10 feet away—thanks to advancements in capture technology—even while a user is moving or wearing contact lenses or glasses (Table 10.1).[9]

[6] Barkleys, "Banking on the Power of Speech", from Barkley's website, https://wealth.barclays.com/en_gb/home/international-banking/insight-research/manage-your-money/banking-on-the-power-of-speech.html

[7] Danielle Beurteaux, "Voice Recognition: Is This The New Killer App?", Digitalist, July 15, 2015, http://www.digitalistmag.com/innovation/2015/07/15/100-million-sound-like-03070180

[8] Chirchi, Waghmare, and Chirchi, "Iris Biometric Recognition for Person Identification in Security Systems", International Journal of Computer Applications, Volume 24– No.9, June 2011, http://www.ijcaonline.org/volume24/number9/pxc3874002.pdf

[9] Mark A. Clifton, "How Hollywood Gets Biometrics Wrong", Sep. 18, 2013, SRI International, https://www.sri.com/blog/how-hollywood-get-biometrics-wrong#sthash.6t9jogwq.dpuf

Table 10.1 Comparison of authentication technologies

	Mis-identification	Security	Application
Iris recognition	1/120,000	High	High security facilities
Finger printing	1/1000	Medium	Universal
Hand shape size	1/700	Low	Low- security facilities
Facial recognition	1/100	Low	Low- security facilities
Signature	1/100	Low	Low- security
Voice printing	1/30	Low	Telephone service

10.3 Cognitive Interaction

Both consumer and organizational communications are full of expression. These communications use a wide mix of communication media, including voice and gestures. Any modern day meeting in a conference room requires white boards, where participants collaboratively draw to share their mental models. Music, pictures, and videos are often used to enhance the expression. Imagine a bunch of consultants, or technical architects in a room. There is always a mad dash to grab attention, and expert collaborators do that skillfully, using gestures to indicate they have something to say, often using unconventional means. People use various ways to attract attention, clapping their hands, thumping feet on the floor, banging their fists on the table, raising their voice, or by threatening to leave the room. They also team up with others to marshal an agenda and may use collaborative actions like looking at each other, winking, or conducting high-fives. Someone takes charge and establishes the agenda. Note-takers busily write down everything and experts take turns to pitch new ideas. What if cognitive things were to be introduced as note-takers, organizers, participants, or experts? In order to support, collaborate, or dominate a conversation, these cognitive things must use a variety of communication techniques, master comprehension of the conversation flow, and be able to proactively grab control wherever needed. This section will discuss communication media and dialog management capabilities of cognitive things and uses illustrative examples to show how far the real world "terminator" is able to converse and collaborate with its human counterparts. In order to participate in human conversations, robots are beginning to use a number of communication media and cognitive capabilities. Let us explore a couple of these cognitive capabilities, which we take for granted as humans, and must now train the Cognitive Things, so they can appropriately interact with humans.

Vision The Cognitive Things use their video cameras and apply visual analysis to establish vision as a cognitive capability. They should be able to use their vision to differentiate between people, animals, and stationary objects, count the number of people in a meeting room, their relative location, their facial features for recognition, as well as emotions expressed by them. Cognitive assistants supporting insurance call centers should recognize car models, read VIN diagrams from photos, and assess damage to a car in recording an accident. Autonomous vehicles driving on streets recognize road and traffic patterns, movement of other cars around them, and especially, cars moving in their path including in blind spots.

Motion Detection and Response The next cognitive capability is around motion. Humans interact by walking towards each other, shaking hands by aligning their movement to others, initiating and receiving a hug, or synchronizing hand movements to conduct a high-five with another person. In an interaction with a human, the cognitive thing requires these cognitive capabilities to express a friendly gesture, such as a hug. It requires cognitive ability to comprehend the reasons for someone else's motions, and to respond to those movements. The cognitive thing also requires both the ability to detect motion and to be able to avoid toppling over or running over objects around them. It takes years for humans to learn how to approach another person, shake hands, avoid obstacles, and lead and follow in Waltz dancing. Robots have a long way to go before they become expert ballroom dancers.

Listening The cognitive thing also requires hearing and listening capabilities to understand and follow conversations, detect human emotion, and comprehend what is being said. Listening skills require many levels of cognition. First, the cognitive thing must be able to differentiate words and sentences from utterances, which is hard because human speech varies enormously in its transition from one word or sentence to another, and people use different accents based on their background. They must then sense the language or languages being spoken. In most technical speak, humans introduce many terms way beyond the standard dictionary. A cognitive thing must be able to apply ontology to understand the technical jargon added to the standard language. Then comes the hard task of parsing a sentence and making sense out of sentences to comprehend meaning. Most smartphones today come with assistants capable of helping their owners find names and phone numbers in the contact list, setting up appointments, taking notes, and conducting searches on apps to find places or to check the weather. Listening to many masters and making sense of all the input is harder, and it is even more difficult to comprehend young children, who have their own words and accents. It is amazing to watch Cognitoys' ability to comprehend utterances from children and engage in discussions.

Dialog The automation era gave us a new and non-intuitive way to deal with interactions—using hierarchical menus. These menus came in many shapes and were shamelessly deployed everywhere, including interactive voice response units (IVRs) where the human was expected to remember the menu hierarchy and its traversal, or in apps where the small user screen made it hard to see and follow the menu. Consumers responded with ferocity, by learning how to hit "0" to opt out of the IVR to connect with a human agent. Cognitive dialog puts an end to this atrocity. It brings a virtual chat assistant capable of presenting a single interface through which the user can express a command while the screen can be reconfigured dynamically to deal with the next conversation. A good dialog management retains the agenda, allows context switching and effectively uses menus and buttons to provide alternatives to typing or speaking.

I was searching for a cognitive thing that would exhibit all the cognitive capabilities described above. I had the pleasure of meeting Pepper, a conversational cognitive robot created by Softbank Robotics. It was a delight to watch Pepper, as he dexterously covered each of the cognitive capabilities described above. Despite his tiny size, Pepper was loaded with sensors to help with his impeccable performance. He was able to recognize voices and authenticate speakers. He was able to directionally follow and turn around as the speakers moved around him. He used a large number of motors to perform a rhythmic dance without toppling over. He was able to hold conversation and switch topics using a human-like fluency not often seen in many primitive interactive voice response units.

Pepper uses four directional microphones located on his head for interactions. These microphones enable him to detect where sounds are coming from, while also allowing him to authenticate the individual as well as identify emotions. Pepper is able to function in complex environments thanks to his 3D camera and two HD cameras that enable him to identify movements and recognize facial emotions. If you are happy, Pepper will share your joy, and if you are sad, Pepper will comfort you. Pepper uses his tablet to help you make choices but also to express his own emotions. This constant dialog between perception, adaptation, learning, and choice is the result of what is known as the emotion engine. Thanks to his anti-collision system, Pepper detects both people and obstacles in order to reduce the risk of unexpected collisions. Pepper is also able to maintain his balance, which stops him from falling if somebody knocks him over. His three multi-directional wheels enable him to move around freely through 360°, at a maximum speed of

3 km/hour using no less than 20 engines (head, arms, back) to control his movements with great precision.[10]

A number of YouTube videos have been created to demonstrate Pepper's multi-faceted capabilities. Animo has applied its voice recognition technology for Pepper's voice authentication.[11] In an interview with CNBC at Consumer Electronics, Rodolphe Gelin, Chief of Innovation for Aldebaran introduced Pepper and used him as a co-speaker in the interview. Pepper was able to follow and lead conversation, express emotions, and respond to friendly gestures from the CNBC reporter.[12]

Automobiles are rapidly gaining interaction capabilities. My 2015 Acura TSX is capable of spotting and alerting me when there are cars in the blind spots. It keeps me in my lane through automatic lane correction by gentling nudging my steering wheel. However, the manufacturer discourages hands-free driving. An advanced version of this car can follow the car in front at a safe distance. Acura does all this using a series of cameras attached to the car. Many cars are offering improved cognitive interface for infotainment. For example, Volkswagen is offering advanced capabilities in its e-Golf Touch version Volkswagen, which is about to be introduced to the market. As you might guess from the name, it includes a 9.2-inch touchscreen infotainment system. The system can also take input via natural-feeling voice interactions as well as gestures. It also sports a fully digital instrument cluster just like the one debuted by the company's Audi subsidiary in the TT roadster in mid-2015. If you thought the e-Golf Touch's digital displays were advanced, BUDD-e goes for a total of three screens, surrounding the driver, where features float on and off the screen as you need them. Volkswagen demonstrated early-stage apps where BUDD-e could hook into your smart-home systems to coordinate when things turn on or off depending upon your location and context (will you be at work for 8 hours or at the grocery store for one hour, for example). Speaking of groceries, it could hook into your smart refrigerator and remind you to pick up more beer while you're out.[13]

---

[10] From Aldebaran's official website, "Cool Robots….Pepper", https://www.aldebaran.com/en/cool-robots/pepper/find-out-more-about-pepper

[11] Animo YouTube post, "Voice Technology Library for Pepper, Voice Authentication Using VoiceTagging", Animo, Nov. 29, 2015, https://youtu.be/g-wto0vFVsk

[12] CNBC, "Robots are taking over CES", CNBC Video Library on YouTube, January 7, 2016, https://youtu.be/nQFgGS8AAN0

[13] Melissa, Riofrio, "How VW is charging forward with its e-Golf Touch and BUDD-e electric cars", PC World, Jan. 6 2016, http://www.pcworld.com/article/3019578/car-tech/how-vw-is-charging-forward-with-its-e-golf-touch-and-budd-e-electric-cars.html

10.4 Emotions, Creativity, and Hidden Meanings

Early science fiction portrayed cognitive things as emotionless agents who were impervious to human emotion. Imagine a society of 6 billion humans and 25 billion cognitive things, where the cognitive things brought the humans to a regimental rational structure. The automation era failed to convert the human race into an emotionless society. Despite all the advances in call center automation technologies, consumers opted out of rigid menus and emotionless digital agents. As stated by my revered professor, Herbert Simon, who received the Nobel Prize in Economics for this major discovery—humans are irrational. We have a creative and emotional mindset, which may sometimes be hidden, as we project a calm and calculated image to the world. We fall in love with ideas, and think creatively outside the box. If the cognitive era is to succeed, it will be because it learns to deal with human emotions and creativity (Cartoon 10.2).

In a typical discussion with a call center, consumers do not always express their real intention directly. They assume the call center person is capable of translating their sentences to find the hidden meaning and work through their emotions. If my smart television sound is not operational, I can start with "My TV is giving me a great silent movie era experience," and would not expect the call center representative to thank me for my good feedback and hang up. A casual key word search to detect emotion may interpret the word "great" to assume that I am happy with my TV performance. However, I am exhibiting sarcasm and unless the representative is familiar with his or her grandmother's days of watching silent movies, they are very likely to not understand the context. When consumers talk to each other, especially with service providers, they often use various emotions—frustration, anger, sorrow, or sarcasm, to express their problem with the service. Cognitive things must understand the emotions, empathize with the consumer, get to the root

Cartoon 10.2 Robot with consciousness
(Scott Adams, "Dilbert", November 23, 2015, http://dilbert.com/strip/2015-11-23 reprinted with permission)

cause or the deeper issue, and resolve the problem. Cognitive things must also match and not overdo the emotion. Someone asking a straightforward question and seeking an efficient answer may get equally frustrated with a roundabout response laden with emotional phrases. The response must match the emotional intensity of the customer and seek to dissipate the emotion using empathy.

Emotion can be detected in the way a consumer speaks and the choice of words a consumer uses to express his or her problems. Phonology is the study of physical sounds of a language and how those sounds are uttered in a particular language. In understanding the call center call, the intonation pattern plays a major role. A person who is angry may use the same words as a person who is confused, however, with a different intonation.

Human personality has a strong impact on how we interact and relate to others. In order to improve interactions, should the cognitive things gauge the personality and appropriately choose the style of interaction? Good servers in a restaurant can increase the amount of tips they receive by adapting their style of conversation to the person they are serving. How would a cognitive thing gauge the personality of an individual they are dealing with? One of the ways to gauge personality is by analyzing the choice of words. By analyzing spoken or written sentences from a person, one can create a personality graph. I used the Watson Personality Insights software in IBM's Bluemix to analyze some of my writing (you can try the service on your text samples using IBM Bluemix app).[14] I am illustrating here my personality as exhibited in my mother's biography that was authored by me (see Fig. 10.2). The personality shown for my business writing (for example the manuscript for this book) is somewhat different. This is an expected outcome as my business writing is likely to reflect more of my work persona, while personal writing reflects my non-work persona.

Personality Insights selects psychographic dimensions based on words chosen by the author in comparison with a baseline. Presumably, my personal writing shows a choice of words that reflects my outgoing and adventurous personality.

Cognitive agents are effective when they show pleasant personalities. As I watched Pepper, I wanted to take one home because Pepper exhibits a nice face with a pleasant personality. Increasingly, designers are engraving positive emotions in the physical design of things. While at home or at the office, take a closer look at the products around you—the desk, the lights, or the doors—you would

[14] Bluemix, "Personality Insights", IBM Published Software as a Service, Feb. 17, 2016, https://personality-insights-livedemo.mybluemix.net

Data Behind Your Personality

Name	Value
Big 5	
Openness	**98%**
Adventurousness	94%
Artistic interests	26%
Emotionality	5%
Imagination	96%
Intellect	100%
Authority-challenging	97%
Conscientiousness	**82%**
Achievement striving	84%
Cautiousness	90%
Dutifulness	28%
Orderliness	14%
Self-discipline	54%
Self-efficacy	92%
Extraversion	**8%**
Activity level	15%
Assertiveness	29%
Cheerfulness	2%
Excitement-seeking	2%
Outgoing	6%

Visualization of Personality Data

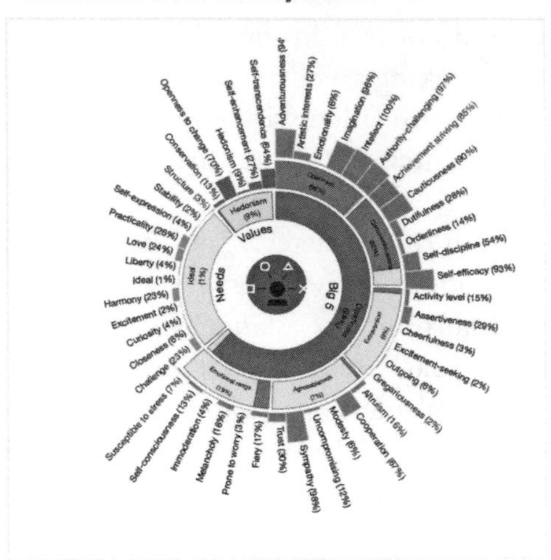

Fig. 10.2 Personality Insights

find an intriguing resemblance between these inanimate objects and human features such as an eye or a mouth, or as a much more complex human emotion as represented in a smile. We might not make our purchasing decisions based on how we perceive products' features, but may unconsciously harbor a deeper feeling based on that object's resemblance to a human feature. Product design encourages the elicitation of various stimuli which when combined contribute to the perceived emotional comprehensions of what the product means and what it stands for. Customers usually prefer familiar features that elicit feelings of pleasure —the features of arched eyes or an upturned mouth translate into feelings of friendliness, which connotes with pleasure and is ultimately preferred over other product facial expressions.[15]

10.5 Negotiation / Disambiguation

Let me refer back to a customer interaction scenario from Chapter 4 (also shown in Fig. 4.3). A customer uses an app to communicate with the wireless service provider. "When is my contract ending?" The app powered by a cognitive brain may consider responding with the date on which the current contract is ending. However, as it digs into its data sources, it finds the customer has recently moved

[15] Dr. Islam Gouda, "Product Anthropomorphism: The Personification of Design", Dec. 21, 2015, DigitalDoughnut,http://www.digitaldoughnut.com/blog/blog/product-anthropomorphism-the-personification-of-design-

and is getting terrible network quality at his new home. The app now has new information not shared by the customer, but that could be very pertinent to the discussion as the customer is possibly calling to help weigh up options to disconnect and transition to a competing provider. The wireless provider may have a Wi-Fi offering to boost signal strength in a hard-to-reach area. Instead of losing the customer, the app can convert this into an opportunity to sell a new product.

This scenario depicts how communication between two parties involves uncovering new facts and constraints not stated upfront. A key aspect of cooperative negotiation is in our ability to share goals, constraints, and options among negotiating parties and coming up with a win-win solution that benefits all parties. A proper negotiation would involve responding to the questions, and carefully easing into follow-up questions without sounding too invasive. After all, the phone company can infer a move to a new dwelling, but the customer may not be ready to tell the world he has moved into a girlfriend's apartment to live. An incorrectly placed question may make it harder to open up the conversation to get to a good outcome.

Sales negotiations often require multiple considerations. In a classic sales situation, a seller must listen to the customer and understand their needs before selling a product. The seller also has some campaigns to run and has an objective to maximize the sale of promoted items. Thus, the tradeoff is between proactive and reactive selling. Good sellers are very good at listening to their customers and understanding their customers' needs. However, they can inject the promotional items at an appropriate time to accelerate the sale. The promotion may be the end goal, but without trust between buyer and seller, the promotion may not be as effective. Complex cooperative negotiation brings elements of timing, dialog, and trust into the conversation, thereby interweaving an orchestration of dialog that is purposeful and directed without pressurizing the buyer.

10.6 Chapter Summary

This chapter has covered many aspects of a human interaction with a cognitive thing. I started with an example of a children's toy, which interacts with small children and evolves with their likes and interests. Robots offer a mechanism to personify Cognitive Things. It is important for the cognitive thing to use proper authentication mechanisms to secure and interact with authorized humans only. We considered the many ways of authentication, other than user ID and passwords. Mechanisms such as voice printing, facial recognition, fingerprints, and iris recognition provide more sophisticated ways of authentication, which also do not require memorizing long and obscure passwords. Using Pepper as an example, a number of cognitive interactions were introduced, including vision, listening, sensing body movements, and dialog. The next topic was a study of tone and psychological traits. By studying voice intonations and choice of words, a cognitive thing can infer customer mood and psychology. Lastly, we saw how Cognitive Things could negotiate and disambiguate conversations, as they interact and mix proactive styles of dialog with reactive questions and answers.

Interaction with humans is the biggest test for machines in their ability to prove their usability. Unlike the automation era, Cognitive Things have the potential to offer a kinder, gentler style of interaction, thereby making it easier for humans to accept machines as cognitive cohorts. In addition to supporting a man–machine interface, Cognitive Things can also assist humans in human-to-human interactions. These interactions will be discussed in the next chapter.

11

Assisting in Human Communications

11.1 Introduction

Four friends are trying to organize a dinner together.

Can we meet for dinner?
Good idea, when?
Where would you like to meet?
What type of food do you like?
Which restaurant would be the most convenient for everyone?
Which day is the most convenient?
Do you need a ride?

Or, they may pick what they did last time, even though everyone complained about the restaurant and food choice. Is there an alternative to endless texts, calls, and emails where a cognitive assistant could sort out options, preferences, constraints, and give a set of options rank ordered by individual preferences?

The corporate equivalent of this scenario is even more prevalent as well as more painful. In order for a couple of busy executives to meet, they may assign the task to their personal assistants, who would then go through their calendars, meticulously choosing alternatives that would meet everyone's calendar constraints, bumping in the process other less important activities and meetings in the process.

Now imagine a set of corporate lawyers working through a joint agreement across a set of organizations, or a set of doctors, including a general practitioner and a group of specialists deciding on a critical patient care. In a typical

A. Sathi, *Cognitive (Internet of) Things*,
DOI 10.1057/978-1-137-59466-2_11

retail store conversation between a sales person and a customer, a consultative discussion takes place where the sales person learns about customer needs, and the customer learns about products being sold. Can we use cognitive assistants to help and support these human interactions and how would they organize themselves and support the human interactions?

How would the cognitive assistant or assistants support a human-to-human interaction? Is this a centralized cognitive agent, or one associated with each individual? Would they first talk among themselves to eliminate many alternatives and present the final few to the humans for the final pick? How would they set themselves up to know all the hard and soft constraints and preferences? Would they go through a laborious process of constraint collection that may take as much time as the negotiation itself? Would they make a choice that is convenient for everyone, or would they give more emphasis to the higher ranking or the most bottlenecked person? Should they develop a hierarchy mirroring corporate organizations structure, and assigne goals and objectives reflecting their organizational goals?

Chapter 9 covered machine-to-machine interactions and showed how cognitive techniques can be used to communicate with each other. In Chapter 10, I discussed interaction between Cognitive Things and humans. In this chapter, I will now extend the discussion to human communication, showing how technology is graduating to intelligent agents for gathering facts, creating alternatives, evaluating and eliminating less worthy options, resolving constraints, and completing transactions. This chapter covers four major topics. First, we look at information discovery, or how Cognitive Things will collect and sort structured and unstructured information in support of a human interaction. Second, alternative generation and prioritization are covered. In order to improve the quality and productivity of human communication, machines may do a lot more than supporting and organizing support data. They may use this data to generate alternatives, evaluate alternatives, and look for ways to improve them. Third, we will consider how this information will be used to support human conversation. Will Cognitive Things listen to the conversation and organize the material, or take an active role in conducting part of the communication? Fourth, we see how the intelligence can be organized using different organizational patterns, covering both centralized and distributed organizations, depending on the requirements for a specific communication.

11.2 Information Integration and Discovery

Let us start with the conversation among friends as described in the previous section. Most likely, this conversation took place on their smartphones, and they possibly used either text messaging, voice conversations, or private messaging in a social conversation app like Facebook messenger. In any case, the conversation most likely was available to the smartphone to listen in on and interpret, obviously with the appropriate permission from the phone owner. In a typical human-to-human conversation today, the Internet of Things are passively engaged, doing a bulk of automation era tasks. In this case, these devices may be engaged in message composition, transmission, storage, and retrieval. The users may be using smartphone apps to locate restaurants, look for their ratings and menu choices. The users may also be relying on human communication to assess distances, availability, and preferences. Although the information is digitally created and available to the smartphone, it requires collaboration across many humans to make it visible to everyone. Can these devices actively participate in information integration and discovery?

For voice conversations, the first task is to convert the speech into text, so that the machines can interpret the conversations. As introduced in Chapter 6, listening to speech is an important observation skill for Cognitive Things. Most smartphones today are able to take voice input and convert it into text. Depending on the situation, the language may also require translation, if some of the conversation happens in other languages.

Once the conversation is available in the form of text, it can be interpreted using natural language classifiers, which perform text analytics. These classifiers categorize raw sentences into intents, thereby assigning meaning to the sentence and interpreting the subject of the conversation. For example, the classifier can detect many ways in which a dinner meeting request is stated. Whether a person types "Shall we meet for dinner?" or "Are you available this evening for a meal", both of these sentences represent an intent to meet for dinner. Good classifiers also adjust for spelling changes, and can filter out chit-chat topics, staying focused on the real intent.

Often, context plays a major role in interpreting a sentence and can be used as a means to decipher the meaning. It can connect pronouns, relating a pronoun reference to a noun in a previous sentence. Also, if I state "where" in response to an invitation to dinner, the "where" must be connected as a request for further elaboration to specify location in relation to the dinner invitation. Most of the automation era software forced explicit conversations or fully spelling out the statements, making it difficult for the software to interpret human conversations. As the technology for cognitive interpretation improves, it becomes more efficient at filling in the gaps and inserting hidden facts.

My smartphone is typically with me and is able to track my location. The location information can also be aggregated to establish my work location, home location, and other hangouts. The location data can also be used to reveal the route I typically take for traveling to work, how often I walk, drive, or take public transport, and where I stop in between for a cup of coffee. With four friends interested in meeting for dinner, this information is sizable, and daunting for human cognition, but this is where Cognitive Things excel the most. Using information from the last several weeks, they can easily tear through a sizable amount of data to locate the best geographical area to meet.

Another interesting aspect is privacy. Would you feel comfortable sharing your typical hangouts and commute with friends? Maybe not, especially, if you are dating a new partner and are not yet ready to declare this to your friends. However, you may be content to share with a cognitive dinner reservation app, which can use the raw data to establish a convenient location to your friends, without disclosing all of your hangout.

Many more disparate pieces of information may require further integration. Software is able to connect the dots a lot more efficiently with past conversations, likes and dislikes, and information from others on restaurants choices and ratings. With pervasive use of smartphones for social activities, the phones have all this information trapped in many apps used by its users, waiting to be integrated and used for something worthwhile like choosing the best restaurant for a group of friends. The local restaurants may be more than willing to monetize this integration effort as it offers them an opportunity to sell a "dinner for four".

11.3 Alternative Generation and Prioritization

So far, the software is able to listen to the conversation, and identify intent to meet for dinner, and is also able to discover, collate, and integrate location, and restaurant data to organize a colossal amount of information. A good crawl solution may just stop here, offer the results to its owners, and let them do the rest. However, a cognitive solution can also generate and prioritize alternatives. To do so, it now must gather some more background information and conduct a series of cognitive tasks.

The original discussion started with one friend asking "Can we meet for dinner?" Was "dinner" in this conversation the real goal and hard constraint, or was the real goal to "meet" and it could have been a dinner or Sunday brunch. In any human negotiation, there are goals and constraints. Some constraints are hard. "I cannot meet for dinner on Wednesday as I have a work deadline to meet for Thursday". Some are soft—for example, "meeting for dinner would be preferred, but lunch may be an option if a common dinner venue is hard to find". How would Cognitive Things identify and use goals and constraints, to generate and evaluate alternatives?

Solutions can be generated using constraint-directed reasoning. A solution space depicts all the values a decision may take. For example, a request to meet for dinner among friends assumes a dinner in the next several weeks, and not within months. Using constraints, a cognitive system can narrow down the options. While there are a finite number of dates available, many of them may need to be eliminated because one of the friends cannot make it. If one friend is terribly busy, or is traveling four days a week on consulting assignments, he is much more constrained and his calendar is the bottleneck, so it should be consulted first. After applying his constraints, the solution space may be much narrower, only leaving Friday, Saturday and Sunday in the next four weeks.

In one of my cherished conversations with Herb Simon, he explained to me the difference between decision-making and problem-solving. The automation-era computing was getting better at decision-making. If all goals and constraints were fully specified, a mathematical programming solution could be applied to generate the optimal solution, by working on the search space and enumerating many alternatives. In a typical decision-making task, all the variables and their values are known. The decision-maker has a narrower objective of finding a set of values, which optimize the goals of the enterprise it is serving. However, problem-solving includes a creative element. The variables and their values are no longer given. It may be possible to generate new alternatives that no one considered. Some constraints are preferences, which are nice to have, but can be relaxed if needed. All efficient executive assistants are masters at problem-solving. To them, nothing is black-and-white, and anything can be changed for a high priority request. In response to the dinner request, by introducing brunch as an option, the solution space is expanded substantially. A semantic model can generate new alternatives, which are close to the current alternatives. Any one with common sense knows "brunch" and "dinner" are good excuses for friends to meet and that "brunch" can be substituted for "dinner" if a suitable "dinner" date cannot be arranged.

The next step is a tradeoff among available options. Alternatives in my example include dates, meeting times, restaurant choices, and ride options. The choice is based on menu, restaurant quality, convenience, parking options, likes and dislikes, past meeting venues and variety seeking, and so on. A tradeoff engine can organize these alternatives, rank them against goals and constraints, and eliminate poor alternatives before the friends are given the final choice of two or three with all the pros and cons.

11.4 Conversation Assistance

If a cognitive calendar event manager can perform all these steps, how is it made visible to the friends? If they were sending messages to each other, would the app seize control, offer assistance, make helpful hints, or reactively manage the information? Despite its ability to help, its level of assistance must be gauged based on human preference. The support could be very reactive, where the cognitive thing offers assistance when asked. It could be very proactive, where the cognitive thing offers advice in the midst of a conversation.

I will use an analogy from sports television coverage to demonstrate how Cognitive Things may support a live discussion among friends. I have always been fascinated by the way a sports television production is able to use a team of experts to cover a live event and keep us engaged as an audience, using a team of experts. The entire session proceeds like clockwork. It is almost like watching a movie; except, the movie is playing live with just a small time buffer to deal with catastrophic events (like wardrobe malfunctions!).

As the game progresses, the commentators use their subject knowledge to observe the game, prioritize areas of focus, and make judgments about good or bad plays. The role of the director is to align a large volume of data, synthesize the events into meaningful insight, and direct the commentators to specific focus areas. It includes replays of moves to focus on something the audience may have missed, statistics about the pace of the game, or details about the players. At the same time, statisticians and editors are working to discover and organize past information, some of which is structured (for example, the number of double faults in tennis, or how much time the ball was controlled by one side in American football). However, other information that is being organized is unstructured, such as an instant replay, where the person editing the information has to make decisions about when to start, how much to replay, and where to make annotations on the screen to provide focus for the audience.

Human-to-human communication can be supported in a similar fashion. Commentators talking to the audience represent the human communication. In the case of sports commentary, the support is increasingly provided by software; for example, to maintain statistics, measure the speed of a serve in a tennis game, or to record and replay the game. The same type of support can be provided in any other human communication. In the case of a sales conversation between a customer and a call center agent, the cognitive thing can observe the conversation, bring associated facts from backend systems, interpret the questions, integrate them with marketing campaigns, and make offers that are best suited to the customer. The call center agent is still in charge and is able to orchestrate the conversation, using the software for a series of support tasks, which normally would have taxed his or her cognitive attention in the midst of a customer conversation.

Cognitive Things may perform a more active role in helping friends find a meeting spot. It is possible for a cognitive meeting planner to act as an intermediary. It may offer messaging using any platform of choice. These messages could be routed through an intelligent agent, who may append expert advice to the messages using a distinguishing mark, thereby providing value over and above a messaging system.

Another way in which Cognitive Things can complement human conversation is where one of the meeting attendees is handicapped and requires aides for the conversations. If a meeting were to include a blind person, how would the blind person understand body language of the people around and whether they are engaged in the conversation? In a heartfelt demonstration that rounded off the Microsoft Build conference keynote, software engineer Saqib Shaikh outlined an ongoing research project that uses machine learning and artificial intelligence to help visually impaired or blind people to better "see" the world around them. The demonstration shows a blind person engaging in a meeting and using a cognitive assistance to describe their reaction to the discussion. Additional scenario shows this person at a restaurant and using cognitive aides to read the menu before ordering food (for details of the video, please see the footnote below[1]). These examples show how Cognitive Things can assist a handicapped person and provide the capabilities to fill in the deficiencies to compensate for deficiencies in loss of vision, speech or hearing. The participation is passive in both of these examples, and is effectively demonstrated by Saqib Shaikh as a way of providing cognitive inputs to his day-to-day activities.

11.5 Organization Communication

So far the discussion was about human-to-human communication at the individual level. The communication is far more complex and cognitively challenging when dealing with organizational communication. Senior management at most large corporations actively uses blogs and videos to

[1] Jason Murdock, "Blind Microsoft engineer unveils AI-powered project that helps him 'see' the world", International Business Times, March 30, 2016, http://www.ibtimes.co.uk/blind-microsoft-engineer-unveils-ai-powered-project-that-helps-him-see-world-1552320

communicate directly with all the employees. In response, they receive a large number of comments from many employees, especially when it comes to major changes to the organization. Often, it is hard for employees to search all the blogs and videos and isolate the relevant ones to act upon. There may be a fair amount of vertical and lateral communication, but is it effectively getting used? This is all within a large organization. How about inter-organizational communication? In dealing with customers, resellers, and suppliers, communication has simply exploded via many communication media. How do I find what is most important for me to read, comprehend, interpret, and act upon?

Social media has created yet another dimension for large scale (mis)communication. For example, a famous celebrity goes to a wireless store and ends up with poor service. Instead of talking to company management, he decides to tweet about the poor performance on Twitter. With a large following and a set of competitors to take advantage of the tweet, the wireless provider faces a large negative amplification of a simple incident, especially if they could not act on it in near real time.

Can Cognitive Things provide some help here? The communication explosion within the corporate environment, as well as across organizations is mostly unstructured text. The automation era excluded analysis and actions associated with this information. However, Cognitive Things can help the employees with a number of cognitive tasks.

As a management blog is created, is it using the right tone and action words? While management signs these blogs, they are often created by staffers. Do they represent the personality of the executive or the staffer writing the blog? The blog could be reshaped to match the personality of the executive and the message. As the blog is communicated, Cognitive Things can monitor the responses and analyze viewership by the impacted employees. As comments are posted, a cognitive thing can organize these comments and summarize them for the busy executive.

When dealing with customers, Cognitive Things can decipher the influence level of an individual, and prioritize communication requiring extra attention as well as engage the appropriate organization and also possibly craft an appropriate response. Cognitive Things can be effective note-takers, observers, and information gatherers in day-to-day meetings, in tricky negotiations, or in group communication. These Cognitive Things do not miss a communication because they are tired, they know how to surface the right issues, and they can turn on a dime in response to changed objective in a restructured organization.

11.6 Chapter Summary

This chapter covered the human-to-human interactions and use of Cognitive Things to support these interactions. We started with a simple example of four friends collaborating with each other to identify a common place and time to meet. Even this simple example illustrates the sizable information available from Cognitive Things, which can be used to help these friends find a convenient geographic location and a time that meets everyone's constraints. As we moved from this simple example to more complex human conversations and then into organizational communication, Cognitive Things could provide a number of activities to support the communication, such as listening to the conversation, identifying intent, gathering facts and supporting data, choosing alternatives, and eliminating based on goals and constraints, thereby narrowing choices to be made.

Humans are creative. They are good negotiators. They can also think outside the box. Cognitive Things can provide the perfect partnership by tirelessly observing, turning over every rock to find supporting evidence, summarizing the information for their human counterparts, and working day and night and in near real time to act before a simple "miss" would get amplified by influencers. The partnership can create a more consistent image for a busy executive, who always publishes the perfect blog with crisp messaging, responds rapidly to feedback, and is able to incorporate the feedback into the next blog. All the hard work in this case comes from the Cognitive Things, which augment and add to the work of the staffers helping the executive.

Cognitive Things can monitor and analyze every conversation employees have with the customers, whether it is a call center representative or a busy doctor dealing with patients. They can look for patterns, identify flaws before it is too late, and provide timely feedback for course correction.

12

Balance of Power and Societal Impacts

12.1 Introduction

As I grew up in 1960s India, I remember a typist school a couple of blocks from my house. I decided to learn how to type, as one of my uncles ran the school and he let me learn without charging any money. While my uncle was glad to see me, he was curious as to why I was interested in learning how to type. "You are good at studies and will most probably become an engineer," he said, "most people come to my school, because they cannot get a professional degree and need to learn typing to get a clerical job." During the 1970s and 1980s, I witnessed a large set of secretarial pools with men and women, each trying their best to show off their typing skills, and their ability to take dictation using shorthand. They would take the morning commute to work, occupy a large number of cubes next to their bosses, and show up with a memo pad and a pencil to take dictation. The typewriters gradually grew their sophistication. They offered capabilities to erase a mistyped letter and store an entire page in its memory, thereby incrementally providing productivity gains for these office workers. In the 1990s, the secretarial pools gradually started to shrink as personal computers showed up in large quantities in the offices. What happened to all those workers, their organizations, and their typing schools?

As human societies grew their technological sophistication, they were able to use technology to improve human productivity. As implementation of technology became widespread, it also had serious implications on society—how we live, interact with each other, organize, and regulate our behaviors. The automation era brought with it large pools of white-collar office workers. Their parents were either blue-collar workers and lived in industrial cities or

A. Sathi, *Cognitive (Internet of) Things*,
DOI 10.1057/978-1-137-59466-2_12

farm workers and lived in rural towns. These office goers worked in downtown offices in large cities and added their share to the morning rush hour commute. Many chose to drive to work, used taxi services, or hired drivers. A large amount of expensive city space was gradually converted into parking lots, mandated by the city planners, and required by the citizens who needed to park their cars.

Cognitive Things bring the next wave of societal changes, which will significantly change markets, governments, and families. As elderly citizens increasingly employ Cognitive Things, they may use fewer social services. As driverless cars increase their share of driving, they may facilitate more effective car-pooling, less need for parking lots, and decreasing auto insurance business. As Cognitive Things start holding personally sensitive information, governments may require access to that data for law enforcement (see the Apple story later in this chapter). As gainfully employed office workers lose their automation era jobs, business organizations and governments may find new ways of using the displaced human capital for new and innovative activities. Will these employees work in cities, or from their homes? How would this transform the cities, the highways, and the suburban residents?

The optimists will paint a more intelligent world with a new set of conveniences and productivity gains. The pessimists will draw horrifying scenarios showing a wider gap between rich and poor or between technical entrepreneurs and routine workers. Privacy advocates will use this opportunity to define new laws that protect the individuals. Educational institutions will need to transform their syllabus to include new educational fields. The impact is staggering and the thinking has barely started. It is hard for anyone to provide the full vision as yet. Let me use the last chapter of this book to pose some important questions and forecast changes, which are very likely by-products of the 25 billion non-human Cognitive Things.

At the beginning of the last century, telephone companies projected a large unfulfilled projected gap of telephone operators who would connect callers in the newly established telecom networks. The gap never materialized because the need was filled by the subsequent automation. Today, organizations are projecting large gaps in data science skills. Will these gaps slow down the deployment of Cognitive Things, or will the Cognitive Things be redesigned to lessen the need for an army of corporate unicorns?

The chapter lays out a couple of important questions. Which org moved to Cognitive Things? As the organizations employ the Cognitive Things, how would it change organizational culture and how would employees, suppliers, and customers redefine their relationships with the organization? How will governments revamp their laws and regulations to safeguard their citizens' interests? How do they balance the rights of one individual against the wider interests of a whole group? As the machines gradually replace office workers, what will be the next frontier to keep humans gainfully employed? Will the machines be organized around the human organization or will there be central pools of Cognitive Things? In each of these areas, I have many questions and insufficient answers, as the thinking has just begun. These are substantive issues to be dealt with by organizations, citizens, and governments worldwide.

12.2 Displacement of Cognitive Jobs

When you need to change the payment mechanism for your insurance, will you search for the answer on your insurance companies website or call them? According to a BBC article, over a million people are employed in call centers in the UK. While it is hard for me to find the total number of call center employees worldwide, their impact on the respective economies are best seen in the busy streets of Manila, where call centers bring in over $25 billion in revenue, making it 10 % of the economy. One Filipino woman, Joahnna Horca, got her social work license only to find that none of the jobs in her industry paid enough for her to support herself. She took a job at a call center working for an American company and now earns $700 per month, which is more than most physicians make in the Philippines.[1] India and Philippines have emerged as major hubs for offshoring call center work. How do customers view these call centers? Many just see them as a nuisance. Time and time again, call centers have been voted one of the most frustrating things to use, with one survey even concluding that calling one is more stressful than going to the dentist.[2] Working at the call centers is equally frustrating. In my studies of call centers in the telecommunications organizations, these jobs have very high attrition rates often approaching 50 %. At the same time, it takes weeks to train a new employee to be a fully productive call center representative.

Cognitive Things offer offshored call centers a natural opportunity or threat, depending on how you view it. In today's world, websites and apps offer natural ways for a consumer to interact with an organization whether dealing with a wireless service provider to pay a bill, or for calling social security administration to change an address. However, a large number of customers use either

[1] BHR Worldwide, "Philippines Becomes Call Center Capital of the World", BHR Worldwide, Feb. 10, 2015, https://www.bhrworldwide.com/philippines-becomes-call-center-capital-of-the-world/

[2] Alex Hudson, "Are call centres the factories of the 21st Century?" BBC Magazine, March 10, 2011, http://www.bbc.com/news/magazine-12691704

chat services or call centers, because the self-service options are difficult to use. Even though a call center may suffer from a wait time, or a chat may require a fair amount of typing and waiting for the chat representative response, organizations continue to see a large number of "unstructured" conversations with their customers. However, in order to offshore these conversations, the call reasons, the recommended dialog flows, and access to organizational systems are fairly well-documented.

Asian manufacturing organizations grew through inexpensive manufacturing in its initial growth period. As its organizations amassed sizable manufacturing business, they graduated towards manufacturing automation and rapidly developed their competitive edge, not through an inexpensive labor force, but through manufacturing equipment and process sophistication. It remains to be seen how the rise of Cognitive Things to replace offshore call centers will reshape outsourcing organizations and economies. As much as manufacturing volumes aided Asia's ability to refine manufacturing expertise, it is an opportunity for outsourcers to take leadership in customer contact automation by gradually improving the sophistication of cognitive customer contact. They could use this opportunity to introduce automated virtual agents for well-documented processes, leaving their human workforce to deal with "open-ended" and "unstructured" conversations.

Is a call center the right way of dealing with customer issues with millennials? Most millennial would rather use their apps to deal with their issues, and conduct a cognitive discussion with a machine rather than wait in line for a human. This self-service option forces the customer to learn more about dealing with their supplier, but in turn also makes the supplier's customer interaction a lot less expensive. Like the telephone operators from the early 1900s who manually connected each call because customers could not dial directly, the call centers filled a void in the late twentieth century when self-service customer interaction processes could not evolve to deal with the variety of issues. As long as the self-service cost reduction is transferred back to the customers in the form of low price, they would gravitate towards cognitive self-service, leading to an increase in programmers who design those self-service systems and a decrease in call center operations.

Machines have several advantages in dealing with call centers. They are tireless, they tend to remember everything stored though with slower performance, and do not show biases. In a typical call center environment, they could cover anywhere from 30 to 70 % of the calls, depending on how much they have been trained. Humans have their own strengths in dealing with call center and chat conversations. They can be creative, can deal with problems never encountered before, and formulate new dialog flows. In the best case, outsourcers can use their best call center employees to deal with the exceptional calls as well as trainers for the machines. Productivity will be improved and will continue to provide employment for the "smarter" call center representatives, but the rest would be replaced with machines.

US labor markets are undergoing important long-term changes. These include:

- The decline of middle-skill occupations, such as manufacturing and production occupations.
- The growth in both high- and low-skill occupations, such as managers and professional occupations at one end, and assisting or caring for others at the other.

Economists have coined the term "job polarization" for this process. As has been argued in the economic literature, the most likely drivers of job polarization are automation and offshoring, as both these forces lower the demand for middle-skill occupations relative to the rest (Fig. 12.1).[3]

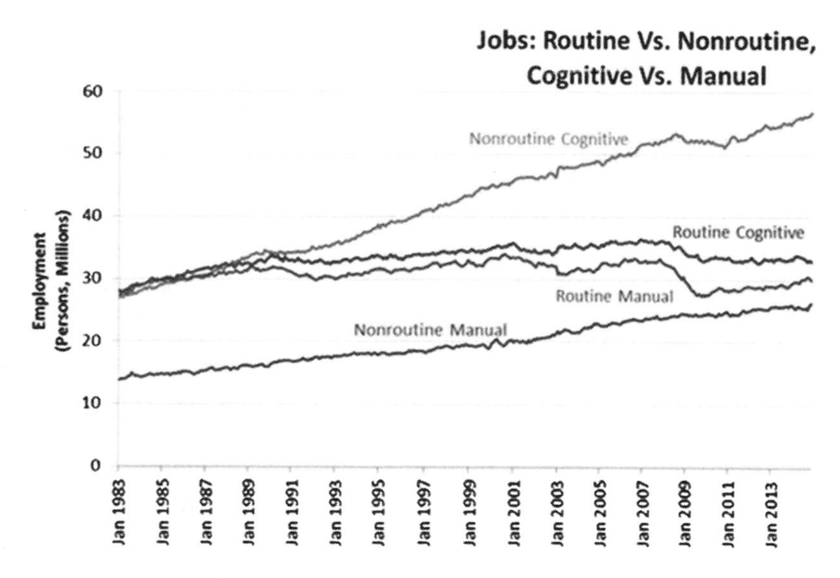

Fig. 12.1 Cognitive employment trends

[3] Maximiliano Dvorkin, Jobs Involving Routine Tasks Aren't Growing, Federal Reserve Bank of St. Louis, Jan. 4, 2016, https://www.stlouisfed.org/on-the-economy/2016/january/jobs-involving-routine-tasks-arent-growing

As routine cognitive jobs are automated, they will result in creating higher unemployment for the labor force they are replacing. The displaced workers must retrain to deal with more nonroutine cognitive tasks, and societies must find ways to encourage new economic opportunities for these displaced workers. In most situations, there are plenty of nonroutine cognitive jobs, which had to be ignored because everyone was busy doing cognitive routine jobs.

12.3 Who Is the Winner?

Each era brings a disruption in the society, tearing apart traditional organizational structures, and making certain skills obsolete. As cars replaced horses, stable owners saw major reduction in their business. As personal computers and smartphones proliferated, they led to the demise of "typing pools" in the organization. In each case, the overall impact of the technological evolution was widely positive and was driven by market forces. Is the Cognitive Era any different?

As stated by Dr. John Kelly, IBM's Senior Vice President for Cognitive Solutions and Research, *"At these moments, it's tempting for people to get caught up in the drama of "man versus machine," perhaps to even worry about the implications of a "loss." But the truth is that whenever AI goes up against the best humans in a field, there is only one real winner—humanity itself. That's the primary reason the world's top computer scientists engage in these games—it helps drive major breakthroughs in technology, which in turn benefits society. The goal isn't necessarily winning the game, but pioneering technology that ultimately advances business, society and individuals."*[4]

Some of the most promising work is already taking place in healthcare, where cognitive computing is identifying insights that can be used by doctors in their fight against deadly and chronic diseases. Cancer centers in New York, North Carolina, India, and Thailand are using these systems to help their doctors in their efforts to provide personalized, evidence-based cancer care tailored to each patient's unique needs. Hospitals in Colorado are helping patients with heart disease adopt—and stick with—heart-healthy behaviors, aiming to help prevent future costly hospital stays. And a new study in California is using technology to determine whether it is possible to predict veterans at risk for long-term Post-Traumatic Stress Disorder (PTSD) and then customize treatment recommendations to help address the devastating condition.

[4] John E Kelly, "IBM's Top Researcher: A Win for Computers Is a Win for Humans", Time, March 21, 2016, http://time.com/4264205/artificial-intelligence-games/

Cognitive Things will significantly increase the productivity of humans and organizations they serve. As much as the automation era provided organizations with a tremendous productivity boost, the cognitive era offers a continued boost in productivity, leading to more output for the same, but retrained labor pool. An interesting aspect of this productivity improvement is in individuals' and organizations' ability to apply their knowledge to new situations. My Under Armour Cognitive Assistant monitors my physical activity, including changes in my weight, food intake and physical activity, and uses a vast pool of resources to give me personalized advice on how to stay focused on my physical activity goals. It also helps me by setting competitions, and social groups for shared activity with people with similar goals. Part of the productivity booster comes from the ability to measure and share the raw data. However, that would inundate me with useless data. It is the ability to analyze the data, compare, build social groups, share insights bringing experts and coaches, where Cognitive Things are beginning to bring a new way to act as my personal trainer and make my physical activity more meaningful. This is a brand new role.

Imagine roads where cars drive in a more orderly fashion—a concept hard to imagine in most cities with poor drivers. I can focus my time on reading emails, talking with coworkers, or conducting a meeting by turning the driver seat around to face the back in a ride-sharing car, because the car is cognitive and knows how to drive itself. It will probably also drive a lot faster on the road, because regulated traffic can move smoothly and may end up stacking cars together. You could lament about the job loss for the Uber driver, the reduction in elderly care workers. Alternatively, you could consider them to be the relics of a past society that lacked Cognitive Things, and use the productivity saving to focus their efforts on nonroutine cognitive activities.

12.4 Regulatory Versus Consumer Privacy

Smartphones have become extensions of the consumers they serve. How do smartphone manufacturers design these phones to protect consumer privacy, while also helping the governments in tracking criminals? Apple computers faced this dilemma in an investigative analysis of terrorist phones in San Bernardino. On December 2, 2015, Syed Rizwan Farook and Tashfeen Malik, a married couple, opened fire at a holiday party at the Inland Regional Center in San Bernardino killing 14 people and injuring 22 people. When the couple were killed in a shootout with police, investigators found an iPhone 5C that Farook had been issued for his job as a county health inspector. Although

FBI agents managed to learn much about Farook and his wife, they wanted to access the phone in the hope that it would help answer outstanding questions, such as whether the killers had accomplices. Concerned that Farook had probably enabled a security feature on the phone that makes it inoperable after 10 failed attempts to enter a secret four-digit security code, agents approached Apple for assistance in getting into the device. Until the correct security code is entered, Apple's encryption software keeps the contents of the phone scrambled. The two sides saw eye-to-eye on a basic idea: the government insisted repeatedly that only Apple's engineers could fashion a way past the security barriers they themselves had built. Apple did not disagree, but it said that being forced to create the work-around amounted to building a master key that, if stolen by hackers, would jeopardize the privacy of all its customers.[5] The specific issue was resolved without a major confrontation, as the FBI found a loophole in the specific version of the iPhone that allowed a hacker to access the data. However, the wider dilemma continues. Should smartphone carry a master key and what happens if the master key falls in the wrong hands?

Manufacturers must create products, which protect their consumers' interests, especially privacy. However, what if these consumers are criminal, and the law requires access to their devices? Master keys are a perfect solution to providing someone with authority access to consumer devices. Smartphones act as cognitive extensions to their brains. They could provide excellent insight into consumers' movement, communication patterns, browsing behavior, and sentiments. Such a rich set of data would make it easy to track a criminal organization. However, the backdoor created for the security of the society can also be used by a criminal to access private data in millions of phones for illegal intents. Clearly the products must be designed to protect the millions of consumers from rogue criminals, but should also support the governments in controlling rogue criminals from misusing the device for criminal activities. While the specific confrontation was avoided because FBI found an alternative way to access the phone, it raised an important policy debate.

Driverless cars will create the ultimate test for policies and how manufacturers and regulators must protect the society in this new cognitive world. Presumably a driverless car will be owned by a leasing company who lends its use to an individual using some form of agreement between the two. The automobile manufacturers today have mechanisms to track and protect a car,

---

[5] Joel Rubin and Paresh Dave, "FBI says it might be able to unlock San Bernardino terrorist's iPhone without Apple's help", LA Times, March 21, 2016, http://www.latimes.com/local/lanow/la-me-ln-feds-looking-at-another-way-to-unlock-terror-attacker-s-iphone-seek-delay-in-hearing-20160321-story.html

and to make it unusable in the case of unauthorized use. The leasing company can protect a car from any unauthorized access to the car, and can possibly remotely control the operation of the car. A traffic authority can sight a car and take control, if needed. So, a person leasing a car can direct the car, but it can be overridden by the leasing company, the manufacturer, or by the traffic authority. However, the overrides must be safely employed to protect the passengers in the car. The overrides are possibly in the form of master keys. As in the dilemma faced by Apple, the automobile master key can be misused by someone intent on carjacking. Policies must dictate how the ownership and control of Cognitive Things would move from one person to another and conditions under which they can be overridden.

12.5 Changing Role of Machines and Humans in Families and Organizations

As 25 billion Cognitive Things are introduced to humankind, they will radically change organizations, families, and society. They will radically alter the traditional roles of caregivers, mothers, doctors, engineers, and everyone else. How would that society behave, and what will be the typical issues? This was my favorite research topic, and I turned to science fiction, movies and television serials to get a glimpse of the future. While most fiction and movies were too futuristic and paranoid about robotic takeover, I did find a couple of interesting glimpses.

Attractive, efficient robots relieve people of the menial tasks of everyday life but problems arise when the androids threaten their human counterparts in *Humans*, a new sci-fi TV series that explores the fascination and fear about technology. *Humans* is not the first show or film to deal with artificial intelligence and robots overtaking humans. But actor William Hurt, who plays scientist Dr. George Millican, said audiences can relate to it because it is not set in a future dystopia. Based on a Swedish TV series, *Humans* takes place in a parallel present in London where highly developed, artificially intelligent servants known as "synths" work in homes and in business. With four interweaving plot lines, it depicts the impact of the latest technology as a suburban family adapts to its robot, detectives investigate synth-related crimes, and a scientist tries to track down renegade androids that can feel emotions. "What this project does is it moves the future into our living room and asks questions in a blunt and organic way about what our visceral reactions would be, which is the best way to entertain your imagination of what this whole issue is about," said Hurt.[6] One of the sub-plots shows a synth supporting an elderly person who needs help with day-to-day chores, such as preparing meals and taking medication. In the story, the elderly struggles with an apathetic synth, who threatens the elderly that any misdirection will be reported to his primary care physician. In another sub-plot a mother questions her role and emotional bond with her daughter, as the synth is able to build a superior emotional relationship with her child.

My wife and I grew up in 1960s India where unified families and household helpers dealt with misdirection by elderlies and competed for children's emotional bonds with mothers. We could relate to the competitive pressure the mother feels in the television serial. Dealing with a Cognitive Thing could be harder, because the synth depicted in that role was perfect. Dino the robot described earlier is already entering the children's lives this year, and will start competing with grandparents as storytellers. How does one compete with a robot with infinite set of stories and jokes, and keep up with todllers' never ending curiosity? As Cognitive Things step into these roles to extend human cognitive activities, we are as likely to depend on them for emotional

Cartoon 12.1 Role of parent vs. driverless car

[6]Adam Bryant, "why Humans is the Scariest Show on TV", TV Guide, June 28 2015, http://www.tvguide.com/news/humans-amc-preview-william-hurt-gemma-chan/

bonding, as we are today with washing machines for cleaning our clothes. It changes the family relationship, but also makes it more secure and easier to manage. With driverless cars, the children may no longer be as dependent on their parents for driving them to school (see Cartoon 12.1), however, the parents may now have more time to relate and interact with the children as opposed to dealing with the task of driving. The elderly person who is likely to fall into depression due to loneliness, will have a constant companion who can change roles, provide a lively environment and also protect the elderly in case of medical emergency. It may not replace a nanny for a long time, but definitely may make it easier for a nanny to take care of two children.

My elderly father was curious about my work. So, I showed him the mechanism by which we can detect if he went to the toilet and did not wash hands after the toilet. He was disgusted with the invasive technology, which could track every flaw and report it to someone. It would be a failure of implementation if the robotic society were emotionless and apathetic, like the synth shown in the television serial. In order for human acceptance, Cognitive Things must represent the value system, the empathy, and respect for individual's decisions.

12.6 Organization Design, Policy Management, Change Management

Human organizations started as hierarchical, but introduced many distributed forms of organizational chains, such as decentralized autonomous divisions, and matrix management to reflect the need for distributed authority. Most successful organizations today are anything but autocratic hierarchies. Formal and informal food chains direct the focus and direction of individuals. Business or project decisions dictate day-to-day relationships for employees. Organizational goals, priorities, and decisions are driven mostly by these forms of organization, while the hierarchies are often the mechanism to aggregate performance for an overall legal entity or its natural divisions, across geographic, functional, or market-driven boundaries.

As Cognitive Things are executers of decisions, and cognitive extensions of their human counterparts, they must reflect the organizational structure of the organizations they represent, and hence must reflect the underlying flow of authority and control. A calendar management administrative assistant makes decisions based on their users, thereby giving higher priorities to revenue generating customers and critical project members. This organizational complexity can be explicitly stated by its users, or inferred from their working relationships by a cognitive assistant. Like a human assistant, the cognitive assistant can grasp changes in organization by reading emails, monitoring conversations, or by sharing "organizational gossip" with other cognitive agents.

Successful organizations often provide tremendous flexibility for their employees, and use policies as a mechanism to maintain the organizational focus, culture, and direction. Business ethics, published organizational culture, and control systems for hiring and employee performance management are used by these organizations to provide management structure. These control systems are often implemented using a set of policies that are communicated and documented using organizational memos, videos, blogs, and monitored via aggregate reporting. It is often hard to steer a large ship, because it requires constant communication across many organizational layers, separated by geographical and political boundaries. In such a setup, Cognitive Things can do a far better job of providing access and management of organizational policies. A set of cognitive agents can provide an efficient system for electronic distribution, tracking, and reporting of policies and related performance. This is far more applicable to federal state and local government, societies, communities of interest, universities, and so on. Policy management requires constant translation of organizational objectives across organizational boundaries. Cognitive Things can do a good job of efficiently extending the human organization. They may also provide means for policing and managing deviant behavior. Driverless cars will do a far better job of sticking to speed limits, and changing their behavior if the traffic authorities decided to reduce traffic speed temporarily to protect its construction workers. They will reduce the need for traffic cops to monitor and control deviant behaviors.

So how would a mayor, a chief operating officer, or president of a university evaluate whether their organizations are functioning properly, or if the change made last month resulted in organizational improvement? Cognitive Things are richer in observations and can be used for impact analysis, experiment design, and organizational optimization. Changes can be made in parts of organization, and based on effectiveness, can be promoted to the rest of the organization. The audit can easily provide a deeper analysis of how the change was received across the organization and where it was the most effective.

12.7 New Skills and Shortage Areas

An organization collects the event data from all the IoTs, synthesizes it with external data, using a set of machine-learning algorithms, the result being a well-organized understanding of the customers, the products, and the human interactions. It sounds like magic! Fortunately, there is a human face to this magic, which makes sense out of all this data. It is termed the "data scientist." As social media and big data companies went after their initial public offerings, media stories catapulted the importance of the data scientist job and the acute shortage of these workers. The data scientist grew from the unappreciated nerd in the back room to a business strategist, a quantitative genius who could consume data for lunch and dinner, and make sense of it. As cognitive business takes shape, the need for data scientist should grow. After all, the data tsunami creates even larger buckets of data. Following my discussions on the profession of data engineers and data scientists in my earlier book,[7] I received many calls from professionals in my network, the aspirant data scientists who wanted to seek advice on how to grow skills to become data scientists. Is this one of the many new skills that will skyrocket like the typists of the 1960s and give a new set of professions to the masses that have been displaced by Cognitive Things?

Cognitive Things are providing organizations with many more observations from the field. They also require improved interface between man and machine, many more robots, and a more human face to the cognitive engineering. This is very likely to fuel the demand for data engineers, data scientists, cognitive engineers, human computer interface designers, robotic engineers, and bioengineers. These are white-collared skills, and require basic skills in statistics, computer science, biology, mechanical engineering and cognitive science. Many large corporations are aggressively hiring in these professions, providing a new campus buzz for educators. As in the previous rounds, the introduction of new technology creates shortages and subsequent demands for new skills, and gradually these skills become more commonplace.

The citizen data scientist movement is one such direction, by which employees in multiple parts of an organization are empowered with the analytics tools and skills to get the answers they need from their data. The movement intends to enable businesspeople to answer 80 percent of their data questions themselves.[8] Companies of all kinds are feeling the acute shortage of trained data scientists today. Even for those lucky enough to snag such a professional, "janitorial" tasks such as data preparation are still taking up an inordinate proportion of those workers' time. Empowering businesspeople to do much of the analysis themselves frees up highly trained data scientists to focus on the things that require their expertise. Part of what's making it possible is the growing set of powerful self-service tools available on the market today,

[7] Arvind Sathi, "Engaging Customers using Big Data", Palgrave Macmillan, 2015, chapter 7, https://www.amazon.com/Engaging-Customers-Using-Big Data/dp/1137386185/

[8] Katherine Noyes, "The rise of the citizen data scientist", Computer World, April 7, 2016, www.computerworld.com/article/3051605/big-data/the-rise-of-the-citizen-data-scientist.html

putting capabilities like artificial intelligence within reach for virtually anyone. In many ways, the citizen data scientist represents an evolution of the traditional business analyst role.

12.8 Chapter Summary

In the beginning of this chapter, we recalled the pool of typists in every 1960s organization. Interestingly, my eleven-year-old niece can type as fast as some of the best typists; granted she is dealing with sophisticated equipment far superior to the mechanical typewriter that jammed after every 10 words if someone typed too fast. Most kids today are far better at expressing themselves using their smartphones or tablets (see Cartoon 12.2). However, the white-collared skill for typing does not carry any extra value, as typing has become the basic skill that everyone has, and no one needs to use shorthand. There is also no need for a separate typing school, though many secondary or high schools have courses on word processors and personal computer-related skill development.

Many of today's cognitive skills are headed the same way. Using a combination of automation, self-service and Cognitive Things, organizations

Cartoon 12.2 Interaction skills and the next generation

of the future will continue to improve their employee productivity thereby reducing the number of employees required for a specific task.

This chapter covered many emerging socio-economic topics, which are as important to the growth of Cognitive Things as the technology itself. Proper process, organization, and policy design at the individual, corporate, and government levels are vital to the continued adaption of Cognitive Things.

Index

A
Adobe, 66
Agarwal, Sunil, ix, 117
Aggregate Knowledge, 66
Alexa, 37
Amazon, 19
  Octocopter, 19
Anderson, John, 38
Apple
  Siri, 37, 104
Apple watch, 2
AT&T
  Digital Life Care, 34

B
Benini, J.P., 38, 138
Blue Kai, 66
Booch, Grady, 7
Business Week, 3

C
Cadillac
  Super Cruise, 16
Cognitoys, 22, 38, 138
CogniToys Dino, 137, 143
Consumer privacy, 167–9
CoreAudience, 66
Cortana, 37

D
da Vinci, Leonardo, 4–5, 21
Driverless cars, 3, 7, 8, 16, 17, 19, 23, 87, 94, 95, 134, 135, 162, 168, 170–2

E
Emergency Care, 34

F
Facebook, 33
  M, 38
Fitbit, 1, 2, 18, 21, 80, 113
Foursquare, 67
Fox, Mark, vii, 97, 133

G
Go Moment
  Ivy, 36

A. Sathi, *Cognitive (Internet of) Things*,
DOI 10.1057/978-1-137-59466-2

Google, 65, 72, 74, 98
  Google Map(s), 103
  Jacquard, 19
  Koala cars, 16

H
Hasse, David, 29, 39
Home security, 15
Horvitz, Eric, 30

I
IBM, vii, viii, ix, 2, 4, 6, 7, 14, 29, 36, 39, 46–8, 52–4, 63, 82, 97, 121, 138, 147, 166
  Almaden Research Lab, 97
  Bluemix, 147
  IBM Research, 97
  Watson, 36, 38, 52, 63, 138

J
Jain, Anshu, 97

K
Kelley III, John, 4, 166
Kenny, David, 63
Knowledge Graph, 97
Kroger, 44
  QueVision, 44

L
Lauren, Ralph, 18
Levi's, 18, 19

M
McKinsey, 3, 4
Microsoft, 99, 103, 157

N
Ninan, Ajit, ix, 117
nPario, 66

O
OBDII, 7, 17
Ontology, 89, 90, 143

S
Set-top box, 65, 88
ShopAdvisor, 32
Slice, 32
Smart television, 13

T
Taco Bell, 32

U
Uber, 36
Under Armour, 2, 18, 22, 112, 167
  HealthBox, 18

W
Waze, 103
Wearable, 18
The Weather Company, 63
Whirlpool, 15, 17